MATHEMATICAL METHODS
IN ENERGY RESEARCH

Proceedings of the Special Year
in the Mathematics of Energy
at the University of Wyoming
July 1982–July 1983

*Sponsored by the University of Wyoming
through the J. E. Warren Endowment,
the Mobil Foundation,
and the National Science Foundation.*

Mathematical Methods in Energy Research

edited by
Kenneth I. Gross
University of Wyoming

siam Philadelphia/1984

Library of Congress Catalog Card Number: 84-52185
ISBN: 0-89871-199-1

CONTENTS

SPECIAL YEAR IN ENERGY MATHEMATICS

PREFACE

During the Academic Year 1982-83 the Mathematics Department at the University of Wyoming sponsored a special year devoted to the mathematics of energy. In light of the abundance of energy and mineral resources in Wyoming, and the resultant importance of the State in the nation's long-term energy outlook, it seems only natural that one should emphasize on campus the importance of mathematical modelling and analysis in energy exploration and production as well as in the associated environmental concerns. Indeed, besides serving as an introduction for both faculty and students to potential research areas in mathematics, the emphasis on physical applications and origins of problems provides an interdisciplinary forum for faculty and students from many other science and engineering departments.

As evidenced by the list of participants and the short-term visitors in the Special Lecture Series, the Energy Year program stressed the close cooperation between mathematicians in academia and those in industry, as well as interaction with mathematical researchers in other areas of science and engineering.

The major areas discussed in courses, seminars, and the visiting lecture series can be loosely grouped under the following five headings: (1) Mathematical modelling related to hydrocarbon recovery, flow in porous media, combustion and chemical reactors. (2) Mathematical analysis of coupled systems of nonlinear partial differential equations. (3) Numerical analysis of transport-dominated flow in two and three dimensions. (4) Computational algorithm development for large, sparse, nonlinear, nonsymmetric systems of equations. (5) Inverse problems, especially related to reflective seismology and geophysical prospecting.

The papers appearing in this Proceedings are quite varied in nature, some are expansions of topics presented in the Energy Year Lecture Series, and others represent recent research on a subject appropriate to these Proceedings. The reader interested in a more systematic expository development of the general subject matter is referred to the first three volumes of the new SIAM series Frontiers in Applied Mathematics. Volume 1, edited by Dr. Richard Ewing, deals with the mathematics of reservoir simulation; Volume 2, edited by Dr. John Buckmaster, concerns combustion and chemical stability; and Volume 3, edited by Dr. Robert Burridge, treats seismic exploration and inversion. It should be mentioned that these volumes are, in part, also an outgrowth of the Special Energy Year at Wyoming.

It is a pleasure to express our gratitude to all who made this special year possible and contributed to its success, including the participants and visitors from academia and industry who gave generously of their time and effort and to the authors who contributed to this volume. Special thanks are due to several individuals. Dr. Richard E. Ewing, the J. E. Warren Professor of Energy and Environment, organized the scientific program and served as a focus for the year's activities. My colleagues, Professors John George, Eli Isaacson, and Duane Porter provided invaluable help. Their efforts during the year were unflagging. I would like to express appreciation to SIAM for its encouragement and support of these Proceedings and to the Rocky Mountain Mathematics Consortium for sponsoring the two summer schools that inaugurated and capped the Energy Year, respectively. Funding for the Energy Year was generously provided by the Mobil Foundation, the National Science Foundation, and the University of Wyoming.

Finally, to close on a personal note, this Special Year in Mathematics Related to Energy launched a period of expansion and development at the University of Wyoming in applied mathematics and interdisciplinary mathematical research on energy and environment. Whatever success will be achieved is due in large part to Dr. L. Milton Woods, Executive Vice President of Mobil Oil Corporation, whose support has been crucial and inspirational. It is a pleasure to be able to record in a formal way our great appreciation to him.

Kenneth I. Gross
University of Wyoming
June, 1984

LIST OF AUTHORS

MYRON B. ALLEN, III, Department of Mathematics, University of Wyoming, Laramie, Wyoming 82071

OWE AXELSSON, Mathematisch Institute, Katholieke Universiteit, Toernooieveld, The Netherlands

ALAIN BOURGEAT, Centre de Mathématiques, Institut National des Sciences Appliquees de Lyon, Av. A. Einstein, 69621 Villeurbanne, France

ROBERT BURRIDGE, Courant Institute of Mathematical Sciences, 251 Mercer Street, New York, New York 10012

JOHN R. CANNON, Mathematics Department, Washington State University, Pullman, Washington 99163

JAMES C. CAVENDISH, General Motors Research Laboratories, Warren, Michigan 48090

EUSEBIUS DOEDEL, Department of Computer Science, Concordia University, 1455 De Maisonneuve Blvd. West, Montreal, Quebec, H3G 1M8

RICHARD E. EWING, Department of Mathematics, University of Wyoming, Laramie, Wyoming 82071

JOHN H. GEORGE, Department of Mathematics, University of Wyoming, Laramie, Wyoming 82071

ROBERT F. HEINEMANN, Research and Development, Mobil Oil Corporation, DRD, P.O. Box 900, Dallas, Texas 75221

JOSEPH V. KOEBBE, Department of Mathematics, University of Wyoming, Laramie, Wyoming 82071

GOEFFREY S. S. LUDFORD, Mechanics Department, Thurston Hall, Cornell University, Ithaca, New York 14853

GUNTER MEYER, School of Mathematics, Georgia Institute of Technology, Atlanta, Georgia 30332

ASOK K. SEN, Department of Mathematical Sciences, Purdue School of Science at Indianapolis, 1125 East 38th Street, Indianapolis, Indiana 46205

MARY F. WHEELER, Department of Mathematical Sciences, Rice University, Houston, Texas 77001

PART I: NUMERICAL AND ANALYTICAL METHODS IN PARTIAL DIFFERENTIAL EQUATIONS

ON THE NUMERICAL SOLUTION OF CONVECTION DOMINATED CONVECTION-DIFFUSION PROBLEMS

O. AXELSSON*

Abstract.

A modified form using an embedding method simplifies the presentation of the streamlined diffusion finite element method of Hughes and Brooks for the numerical solution of convection dominated convection-diffusion flow problems. The modification also makes it more clear which types of boundary conditions to choose in connection with a method of separating the layer part calculation of the solution. We discuss also more classical methods, such as exponentially upwinded Galerkin methods and a new defect-correction method.

1. Introduction.

The numerical solution of convection dominated convection-diffusion problems is difficult because the solution has layers and a standard Galerkin method may result in severely oscillating approximate solutions, unless the finite element mesh is very fine. Furthermore, even if the exact solution is smooth (i.e. has no layers) such methods do not in general give optimal orders of accuracy in the discretization parameter (h).

The nonphysical oscillations may be damped out by introducing an artificial diffusion term in the discrete problem of O(h) (see for instance [15]). Alternatively, one may use upwinded basisfunctions such as in [17]. In the limit case, and for piecewise linear finite element basisfunctions, this latter method approaches an upwinded difference method. Besides the usual limitation in the order of accuracy, classical artificial diffusion or upwind finite elements have the

*Department of Mathematics, University of Nijmegen, The Netherlands

undesirable property of smearing sharp fronts in the directions perpendicular to the streamlines. This is caused by the fact that the artificial diffusion term acts in all coordinate directions. In some problems, such as miscible displacements problems one encounters also an undesirable grid-orientation effect: the approximate solution may depend heavily on the orientation of the grid, see [8].

By use of an exponentially damped weightfunction used in the layer elements only (see for instance [1]), one may damp oscillations caused by a downstream layer (see below), but the oscillations due to characteristic layers remain. Use of an exponentially damped weightfunction all over the domain may cause numerical instabilities when evaluating the exponential functions. See however the discussion at the end of Section 3 about the use of such a method for the approximate symmetrization of the given problem.

To be more specific we shall consider the numerical solution of the convection-diffusion problem:

(1.1) $L_\varepsilon u = -\underline{\nabla}\cdot(\varepsilon\underline{\nabla} u) + \underline{v}\cdot\underline{\nabla} u + qu = f$ in $\Omega \subset \mathbb{R}^n$, $u = \alpha$ on Γ_-, $u = 0$ on $\Gamma_{D_0} \subset \Gamma_0$, $\underline{\nabla} u\cdot\underline{\hat{n}} = 0$ on $\Gamma_0 \setminus \Gamma_{D_0}$ and boundary conditions on Γ_+ to be specified later. Here some concentration u (of a chemical, of heat, etc) is driven by a velocity field $\underline{v}$ and

$$\Gamma_- = \{\underline{x} \in \partial\Omega;\ \underline{v}\cdot\underline{\hat{n}} < 0\}$$

$$\Gamma_0 = \Gamma_{0,1} \cup \Gamma_{0,2} = \{\underline{x} \in \partial\Omega;\ \underline{v}\cdot\underline{\hat{n}} = 0\}$$

$$\Gamma_+ = \{\underline{x} \in \partial\Omega;\ \underline{v}\cdot\underline{\hat{n}} > 0\}$$

are the inflow, the characteristic and outflow boundary parts, respectively, of $\partial\Omega$ (see figure 1 for an example). ε is a diffusion tensor the components of which are small for convection dominated flows. For most of the discussion there is no limitation in assuming that ε is a scalar parameter.

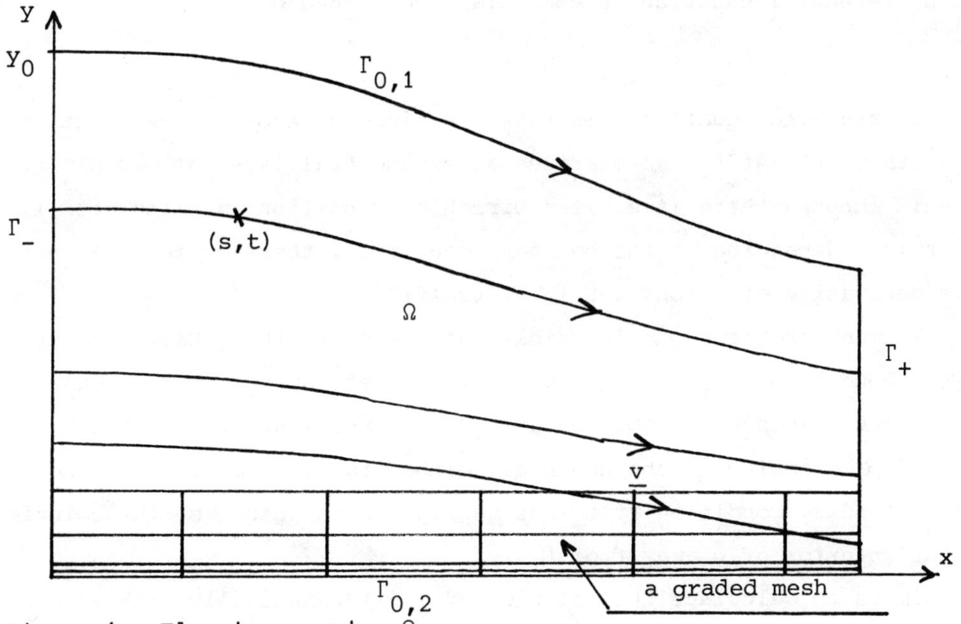

Figure 1. Flow in a region Ω.

In practice, the characteristic boundaries are often physical, like a wall. We shall assume that the vector function $\underline{v}$ is $[C^1(\Omega)]^n$, that $q \in C^1(\bar{\Omega})$ and that $f \in L^2(\Omega)$. We also make the standard assumption that $q - \frac{1}{2}\underline{\nabla}\cdot\underline{v} \geq q_0 \geq 0 \ \forall \underline{x} \in \Omega$. In general $q_0 = 0$ but in some cases we shall assume that $q_0 > 0$. For part of the theoretical analysis we may introduce a new coordinate system (s,t) where t is the parametric representation of Γ_- ($t=y/y_0$ in Fig. 1) and $s = s(t)$ the parametric representation of the vector field line beginning at t on Γ_-. We let

$$\frac{d\underline{x}(s,t)}{ds} = \underline{v}(\underline{x}(s,t)),\ 0 < s < s^*(t),\ \underline{x}(0) = \begin{bmatrix} 0 \\ ty_0 \end{bmatrix} \quad \text{(in Fig. 1)}.$$

The solution of the reduced equation $L_0u = f$ (which is a scalar first order hyperbolic equation) satisfies then

$$\frac{du(s,t)}{ds} + qu(s,t) = f(s,t),\ 0 < s < s^*(t),\ u(0,t) = \alpha(0,ty_0),$$ i.e. for each t, an ordinary differential equation along the characteristic lines. Clearly when ε is small, the solution of (1.1) is essentially governed by the reduced solution and in general the boundary conditions on $\Gamma_0 \cup \Gamma_+$ are not satisfied. Hence there arises layers

along $\Gamma_0 \cup \Gamma_+$.
Along and close to $\Gamma_{0,2}$ (similar considerations are valid for $\Gamma_{0,1}$), the differential equation is essentially governed by

$$L_\varepsilon u \simeq -\varepsilon u_{yy} + v_1 u_x + qu = f,$$

i.e. a parabolic equation. For this one finds a layer in the solution of width $O(\sqrt{\varepsilon})$. At Γ_+ there arises an exponential layer of width $O(\varepsilon)$ if u is imposed there (i.e. if a Dirichlet condition is valid at Γ_+), otherwise, depending on the boundary condition, there arises a layer in some derivative of u (but not in u itself).

In many problems one is mainly interested in the behaviour of the layer along $\Gamma_0 = \Gamma_{0,1} \cup \Gamma_{0,2}$ but not in the one at Γ_+. In such cases we may select a boundary condition at Γ_+ for convenience of smoothness and ease of programming of the numerical method to be used, because the exact boundary condition at Γ_+ has negligible influence on the solution in the interior of Ω away from Γ_+.
Our aim is to calculate the solution of (1.1) numerically with error estimates of high order and valid uniformly in ε. Two difficulties arise then. The first is associated with the lack of regularity of the solution valid uniformly in ε. Typically when there is a layer at Γ_+, the solution $u = u_\varepsilon$ satisfies
$\varepsilon^{3/2}\|u_\varepsilon\|_2 + \varepsilon^{1/2}\|u_\varepsilon\|_1 + \|u_\varepsilon\|_0 \le C[\|f\|_0 + |\alpha|_{\Gamma_-}]$ where $\|\cdot\|_s$ is the Sobolev norm of s'th order and $|\cdot|_{\Gamma_-}$ is the L_2 norm on Γ_-. By considering a one-dimensional problem it follows that this inequality is sharp in the sence that $\|u_\varepsilon\|_2$ diverges as $\varepsilon^{-3/2}$ as $\varepsilon \to 0$. For problems without an outflow layer a somewhat better regularity result may be derived, see e.g. [4].

The second difficulty is due to the lack of coercivity valid uniformly in ε. If $q - \frac{1}{2}\underline{\nabla}\cdot\underline{v} \ge q_0 > 0\ \forall \underline{x} \in \Omega$, we have only
$\int_\Omega (L_\varepsilon u)u d\Omega \ge c[\varepsilon\|u\|_1^2 + \|u\|_0^2]$ and this is the cause of a degraded order of accuracy of the classical Galerkin method (see Section 3) for small values of ε.

To overcome the problem associated with the lack of regularity due to the layers we split the solution in three terms, a smooth one corresponding to Neuman boundary conditions or boundary conditions defined by the reduced equation at $\Gamma_0 \cup \Gamma_+$, a term in the form of a smooth function multiplying an exponential layer function at Γ_+ and a term taking care of a possible layer along Γ_0. In calculating the last term, we think of using a refined (graded) mesh along Γ_0 to resolve this layer (see Fig. 1.). This is not unreasonable, in as much the width of the layer is $O(\sqrt{\varepsilon})$ there. The smooth term may be calculated numerically by any of the methods discussed in Sections 3 and 4. This method of splitting the solution does however not handle possible interior layers. Methods of calculating the layers separately has been considered for one-dimensional problems in [5] and [6].

In order to reduce the non-physical smearing of sharp fronts (such as in an interior layer) in directions perpendicular to the convective field force, Raithby proposed in [19] an upwinded finite difference scheme which however was still only first order accurate but having an artificial diffusion term acting only in the direction of the streamlines. The idea was extended to finite elements by Hughes and Brooks in [11], and was further developed by Hughes and Brooks in a series of papers. The extra diffusion introduced by these so called streamline diffusion methods may be compensated for by a proper use of a Petrov-Galerkin formulation. The mathematical analysis of this method was done by Johnson and Nävert [13] for a problem with constant coefficients. Nävert extended the method in [18] to cover more general stationary and time-dependent problems. In particular he proved discretization error estimates valid in the interior of $\overline{\Omega}$.

In the present paper, we consider only global error estimates and valid uniformly with respect to ε. In section 3 we discuss the classical Galerkin method and in the last section we discuss a modified form of the streamline diffusion method. The modification is done by embedding the solution of the given problem in the solution space of a modified problem of a higher order with an artificial diffusion term acting along the streamlines. Then a classical Galerkin method may be applied on the new equation. For this we prove readily almost optimal orders of convergence of the H^1- and L^2-discretization errors. The advantage of the modified form is that the analysis is somewhat simplified and we see easily which boundary conditions to use in order not to introduce extra layer effects, when we use the method of splitting the layer parts from the solution.

2. Method of separating the layer parts of the solution.

We shall use a modification of a method by Levinson (see e.g. [16]) to split off the layer parts of the solution. We shall assume that $\underline{\nabla}\cdot\underline{v} \le 0$, that ε is constant and that there exists a function g, such that $g = 0$ on Γ_+, $\underline{\nabla}g\cdot\hat{\underline{n}} = -\underline{v}\cdot\hat{\underline{n}}$ on $\Gamma\backslash\Gamma_+$ and that, at least approximately, $\underline{\nabla}g = -\underline{v}$ (see [1], p. 337, for a method to transform the problem to a form where this is satisfied). For ease of exposition, we assume in fact that this is satisfied exactly, i.e. $\underline{v}$ is a potential vector field. Then $\underline{\nabla}^2 g = -\underline{\nabla}\cdot\underline{v} \ge 0$ in Ω and because $\underline{\nabla}g\cdot\hat{\underline{n}} \le 0$ on Γ_+, it follows that $g > 0$ in Ω. Let at first $u = u_\varepsilon^{(0)} + u_\varepsilon^{(1)}$, where

$$L_\varepsilon u_\varepsilon^{(1)} = f,\; u_\varepsilon^{(1)} = \alpha \text{ on } \Gamma_-,\; \underline{\nabla}u_\varepsilon^{(1)}\cdot\hat{\underline{n}} = 0 \text{ on } \Gamma_0$$

and $u_\varepsilon^{(1)}$ satisfies the prescribed boundary conditions for u_ε at Γ_+. Assume in this section that these are also of Dirichlet type, i.e. $u_\varepsilon^{(1)} = \beta$ at Γ_+. For $u_\varepsilon^{(0)}$ we have then

$$L_\varepsilon u_\varepsilon^{(0)} = 0,\; u_\varepsilon^{(0)} = 0 \text{ on } \Gamma_- \cup \Gamma_+,\; \underline{\nabla}u_\varepsilon^{(0)}\cdot\hat{\underline{n}} = 0$$

on $\Gamma_0 \setminus \Gamma_{D_0}$ and $u_\varepsilon^{(0)} = u_\varepsilon - u_\varepsilon^{(1)}$ on Γ_{D_0}. Hence $u_\varepsilon^{(0)}$ contains the layer along the characteristic lines and $u_\varepsilon^{(1)}$ is free of such a layer. We now separate the downstream layer from $u_\varepsilon^{(1)}$,

$$u_\varepsilon^{(1)} = \tilde{u}_\varepsilon + ze^{-g/\varepsilon} \tag{2.1}$$

where g is defined above and where $\tilde{u}_\varepsilon$, z will be defined below. Then

$$L_\varepsilon u_\varepsilon^{(1)} = L_\varepsilon\tilde{u}_\varepsilon + e^{-g/\varepsilon}\{[-\underline{\nabla}(\varepsilon\underline{\nabla}z) + (2\underline{\nabla}g+v)\cdot\underline{\nabla}z + (q+\nabla^2 g)z] - \tag{2.2}$$
$$-\frac{1}{\varepsilon}[\underline{v}\cdot\underline{\nabla}g + |\underline{\nabla}g|^2]z\} = f.$$

We shall let

(2.3) $\quad L_\varepsilon \tilde{u}_\varepsilon = f$ in Ω , $\tilde{u}_\varepsilon = \alpha$ on Γ_- and $\underline{\nabla}\tilde{u}_\varepsilon \cdot \underline{\hat{n}} = 0$ on Γ_0 (i.e. the boundary conditions at $\Gamma_- \cup \Gamma_0$ as for $u_\varepsilon^{(1)}$). At Γ_+ we may choose proper boundary conditions in order to get a smooth solution $\tilde{u}_\varepsilon$ without layers. For simplicity, we take $\underline{\nabla}\tilde{u}_\varepsilon \cdot \underline{\hat{n}} = 0$. Since $\underline{\nabla}^2 g = -\underline{\nabla}\cdot\underline{v}$ it follows from (2.2) that z must satisfy $\underline{\nabla}\cdot(-\varepsilon\underline{\nabla}z) + (2\underline{\nabla}g+\underline{v})\cdot\underline{\nabla}z + (q+\underline{\nabla}^2 g)z = 0$ or

(2.4) $\quad L_\varepsilon^* z = \underline{\nabla}\cdot(-\varepsilon\underline{\nabla}z) - \underline{\nabla}\cdot(\underline{v}z) + qz = 0$ in Ω.

The boundary conditions are

$$z = u_\varepsilon^{(1)} - \tilde{u}_\varepsilon \text{ on } \Gamma_+,\ z = 0 \text{ on } \Gamma_-.$$

Further $\underline{\nabla}z\cdot\underline{\hat{n}} = 0$ on Γ_0 because

$$0 = \underline{\nabla}(u_\varepsilon^{(1)} - \tilde{u}_\varepsilon)\cdot\underline{\hat{n}} = [\underline{\nabla}z\cdot\underline{\hat{n}} - \frac{1}{\varepsilon}\underline{\nabla}g\cdot\underline{\hat{n}}]e^{-g/\varepsilon} = [\underline{\nabla}z\cdot\underline{\hat{n}} + \frac{1}{\varepsilon}\underline{v}\cdot\underline{\hat{n}}]e^{-g/\varepsilon}$$

and, by definition, $\underline{v}\cdot\underline{\hat{n}} = 0$ on Γ_0.

Hence z is the solution of the adjoint operator equation. For this the flow goes in the opposite direction and an exponential layer may now occur at Γ_-.

In the same way as above in (2.1), we may split off the layer part of z, i.e. $z = z^{(0)} = \tilde{z}_\varepsilon + z^{(1)}e^{-g^{(1)}/\varepsilon}$, where $g^{(1)}$ satisfies $\underline{\nabla}^2 g^{(1)} = \underline{\nabla}\cdot\underline{v}$ in Ω, $g^{(1)} = 0$ on Γ_- and $\underline{\nabla}g^{(1)}\cdot\underline{\hat{n}} = \underline{v}\cdot\underline{\hat{n}}$ on $\Gamma\backslash\Gamma_-$.

$\tilde{z}_\varepsilon$ satisfies $L_\varepsilon^*\tilde{z}_\varepsilon = 0$ and $z^{(1)}$ satisfies $L_\varepsilon z^{(1)} = 0$. The process may be repeated until the corrections (the layer terms) are small enough. Note that, because $g > 0$ in the interior of Ω, the correction terms, $ze^{-g/\varepsilon}$ etc. are indeed small in the interior, when ε is small. This implies that we only have to calculate g numerically in the vicinity of Γ_+.

In the above fashion, our problem (1.1) is reduced to solving problems $L_\varepsilon\tilde{u}_\varepsilon = f$, $L_\varepsilon^*\tilde{z}_\varepsilon = 0$ etc., where the solution is smooth (has no layers).

<u>Remark 2.1.</u> Above we have assumed that $\alpha \in C^0(\Gamma_-)$. If α is discontinuous at some point on Γ_-, there arises an interior layer, of the same type as the characteristic boundary layers. Interior layers can't so easily be treated in the above way.

Remark 2.2. An alternative but very similar way of utilizing the exponential function in (2.1), $u_\varepsilon = \tilde{u}_\varepsilon + ze^{-g/\varepsilon}$, is by use of enriched subspaces. Then we add a function $z_h\, e^{-g/\varepsilon}$ to the usual polynomial finite element subspace V_h in order to model the downstream boundary layer. Here z_h is a function in V_h with support only at elements at Γ_+. We may take

$$g(x,y) = v_1(x_0,y_0)(x_0-x) + v_2(x_0,y_0)(y_0-y)$$

where (x,y) is a point on the characteristic line through the point (x_0,y_0) on Γ_+. In practice it is likely that this method is to be recommended over the splitting off method. For an application of this in a one-dimensional problem, see [14].

3. The classical and exponentially upwinded Galerkin method.

In order to find $\tilde{u}_\varepsilon$ (and similarly for z) we consider the variational formulation: Find $u \in \overset{\circ}{H}{}^1(\Omega)$ such that

$$\text{(3.1)}\qquad a(u,\tilde{u}) = \int_\Omega L_\varepsilon u\, \tilde{u}\, d\Omega = \int_\Omega (\varepsilon\underline{\nabla}u\cdot\underline{\nabla}\tilde{u} + \underline{v}\cdot\underline{\nabla}u\, \tilde{u} + qu\tilde{u})\,d\Omega =$$

$$= \int_\Omega f\tilde{u}\,d\Omega \quad \forall\, \tilde{u} \in \overset{\circ}{H}{}^1(\Omega) = \{u \in H^1(\Omega);\ u = 0 \text{ on } \Gamma_- \cup \Gamma_{D_0}\}\ .$$

(For notational simplicity, we have assumed that $\alpha = 0$ on Γ_-.) We assume that

$$\text{(3.2)}\qquad q - \tfrac{1}{2}\underline{\nabla}\cdot\underline{v} \ge q_0,\ x \in \Omega,\ q_0 \text{ a positive constant,}$$

in which case we get by Green's formula and standard inequalities,

$$a(u,u) \ge \|u\|^2_{\varepsilon,1} \equiv \int_\Omega (\varepsilon|\underline{\nabla}u|^2 + q_0u^2)\,d\Omega + \oint_{\Gamma_+} \underline{v}\cdot\hat{\underline{n}}\, u^2\, d\Gamma \quad \forall\, u \in \overset{\circ}{H}{}^1(\Omega).$$

This proves coercivity of the bilinear form and stability in L_2-norm for all ε. If condition (3.2) is not satisfied, a variational formulation with an exponential weight function,

$$\text{(3.3)}\qquad a(u,\tilde{u}) = \int_\Omega L_\varepsilon u\, \tilde{u}\, e^{-g} d\Omega = \int_\Omega f\tilde{u}\, e^{-g} d\Omega \quad \forall\, \tilde{u} \in \overset{\circ}{H}{}^1(\Omega),$$

will yield such a form. Here g is essentially independent on ε (see [1]).

Let $V_h \subset \overset{\circ}{H}{}^1(\Omega)$ be a regular finite element space of piecewise polynomials of degree k, where h is a mesh length parameter and let the Galerkin approximation u_h be defined by

(3.4) $a(u_h,\tilde{u}) = \int_\Omega f\tilde{u}d\Omega \;\forall\; \tilde{u} \in V_h.$

Let u_{I_h} be the interpolant of u on V_h, let $\theta = u_h - u_{I_h}$, and $\eta = u - u_{I_h}$.
Note that $\theta \in V_h$ and that we easily derive estimates of the interpolation error η by use of Taylor expansions of u. We get from (3.1), (3.4)

$$a(\theta,\theta) = a(\eta,\theta).$$

This relation is the basis for the following error estimate.

Theorem 3.1. Assume that $q - \frac{1}{2}\underline{\nabla}\cdot\underline{v} \geq q_0 > 0 \;\forall\; \underline{x} \in \Omega$.
Then the discretization error $u-u_h$ of the Galerkin approximation u_h defined by (3.2) satisfies

(3.5a) $\|u-u_h\|_0 \leq Ch^s\|u\|_{s+1}$, if $0 < \varepsilon \leq h^2$

(3.5b) $\|u-u_h\|_0 \leq C\varepsilon^{-\frac{1}{2}}h^{s+1}\|u\|_{s+1}$, if $h^2 \leq \varepsilon \leq h$,

(3.6a) $\|\underline{\nabla}(u-u_h)\|_0 \leq \varepsilon^{-1}h^{s+1}\|u\|_{s+1}$, if $h^2 \leq \varepsilon \leq h$

(3.6b) $\|\underline{\nabla}(u-u_h)\|_0 \leq Ch^s\|u\|_{s+1}$, if $h \leq \varepsilon$.

Here $0 < s \leq k$.

Proof. (cf. [18]) We have

(3.7) $a(\theta,\theta) \geq \int_\Omega (\varepsilon|\nabla\theta|^2 + q_0\theta^2)d\Omega + \oint_{\Gamma_+} \underline{v}\cdot\hat{\underline{n}}\theta^2 d\Gamma.$

To bound $a(\eta,\theta)$ we use standard inequalities to get

$$a(\eta,\theta) \leq \frac{1}{2}a(\theta,\theta) + C[\int_\Omega \varepsilon|\nabla\eta|^2 d\Omega + \oint_{\Gamma_+} \eta^2 d\Gamma]$$

$$+ C\begin{cases} \int (|\underline{\nabla}\eta|^2+\eta^2)d\Omega, & \text{if } 0 < \varepsilon \leq h^2 \\ \int \varepsilon^{-1}\eta^2 d\Omega, & \text{if } h^2 \leq \varepsilon \end{cases}$$

In the first case, the term $\int \underline{v}\cdot\underline{\nabla}\eta\theta d\Omega$ has been estimated directly, and in the second case we have first used Green's formula,

$$\int \underline{v}\cdot\underline{\nabla}\eta\theta = -\int \eta\underline{v}\cdot\underline{\nabla}\theta - \int \underline{\nabla}\cdot\underline{v}\nu\theta + \oint \underline{v}\cdot\hat{\underline{n}}\,\eta\theta$$

and then estimated

$$|\int \eta \underline{v}\cdot\nabla\theta| \le C\varepsilon^{-1} \int \eta^2 + \frac{1}{2}\varepsilon \int |\underline{\nabla}\theta|^2$$

and the remaining terms by Cauchy-Schwarz' inequality.
Hence

$$a(\theta,\theta) \le C[\varepsilon h^{2s} + h^{2s+2} + \begin{cases} h^{2s} \\ \varepsilon^{-1} h^{2s+2} \end{cases}] \| u \|_{s+1}$$

and by the triangle inequality and (3.7), the Theorem follows.

Remark 3.1. Note that we can't improve these estimates by a duality argument ('Aubin-Nitsche trick'). (3.6b), however, can be used to improve the L_2-norm error estimate when $\varepsilon > h$.
We note that it is only the estimate (3.6b) which is of optimal order. When $\varepsilon \le h^2$, the estimate (3.5a) is one order less than optimal.

More severe however is that there may be no convergence at all. In general this happens when the solution u has a downstream layer. Consider the case k = 1. Then, as was remarked in Section 1, $\| u \|_2 \sim \varepsilon^{-3/2}$, $\varepsilon \to 0$, which means that neither (3.5), nor (3.6) shows any convergence when $\varepsilon \le h$. When $\varepsilon = h$ we get from (3.5b) only $\| u-u_h \| \le C$, i.e. boundedness with respect to h, $h \to 0$. Practical computations show that the Galerkin solution oscillates around the exact solution when this has a layer.

However, due to the splitting made in Section 2, there is no such layer for $\tilde{u}_\varepsilon$ (and z). It follows then that we have convergence. For a recent discussion of regularity of the solution under various boundary conditions, see [4].

Remark 3.2. If we let $h \le c\varepsilon$, c small enough we get a positive scheme for k = 1 and a proper triangulation of Ω. From this we may derive an error estimate $\| u-u_h \| = O(h)$, in the usual manner (see e.g. [15]).

From the above we see that the classical Galerkin method gives nonoptimal error estimates when $\varepsilon < h$ and in general does not converge if the solution has a layer, due to a Dirichlet boundary condition on Γ_+. In the method of artificial diffusion one simply adds a term $-c^{-1}h\underline{\nabla}\cdot\underline{\nabla}u$, c small enough, to $L_\varepsilon u$, which means that the solution is perturbed by an amount O(h) and hence the error is never smaller than O(h). Furthermore, when the solution has a shock wave travelling along a characteristic line, this will be serverely smeared out, which is often undesirable.

In the last section we shall discuss a modification of a recently proposed method which has the following advantages compared to the classical Galerkin method and the method of artificial diffusion.

(i) It converges for all ε.

(ii) The rate of convergence is improved when $\varepsilon < h$.

(iii) It has no smearing effects.

Due to characteristic line layers there may however still appear some oscillations around such layers unless $h \le c\sqrt{\varepsilon}$, c small enough. A method which will globally damp all oscillations is based on the exponentially weighted variational formulation (3.3) with $g := g_\varepsilon = O(\varepsilon^{-1})$. To derive this, consider (1.1), for simplicity with homogeneous boundary conditions and ε constant. We have

$$\int_\Omega L_\varepsilon u\, \tilde{u}\, e^{-g_\varepsilon} d\Omega = \int_\Omega f e^{-g_\varepsilon} d\Omega \quad \forall\, \tilde{u} \in \overset{\circ}{H}{}^1(\Omega)$$

and after some elementary calculations (see [1]) we get

$$a^{(g_\varepsilon)}(u,\tilde{u}) = \int_\Omega [\varepsilon \nabla u \cdot \nabla \tilde{u} + (\underline{v}_\varepsilon \cdot \nabla u + p_\varepsilon u)\tilde{u}] e^{-g_\varepsilon} d\Omega,$$

where $\underline{v}_\varepsilon = \underline{v} - \varepsilon \underline{\nabla} g_\varepsilon$ and $p_\varepsilon = q - \underline{\nabla} \cdot \underline{v}_\varepsilon + \underline{v}_\varepsilon \cdot \underline{\nabla} g_\varepsilon$.

Assume at first that $\varepsilon^{-1}\underline{v}$ is a potential vector field. Then we may choose g_ε such that $\underline{\nabla} g_\varepsilon = \varepsilon^{-1}\underline{v}$, i.e. so that $\underline{v}_\varepsilon \equiv 0$. Hence

$$a^{(g_\varepsilon)}(u,\tilde{u}) = \int_\Omega [\varepsilon \underline{\nabla} u \cdot \underline{\nabla} \tilde{u} + q u \tilde{u}] e^{-g_\varepsilon} d\Omega$$

which is a symmetric bilinear form. Using a Galerkin method

$$a^{(g_\varepsilon)}(u_h,\tilde{u}) = \int_\Omega f e^{-g_\varepsilon} d\Omega \quad \forall\, \tilde{u} \in V_h, \tag{3.8}$$

we may now derive error estimates as for symmetric bilinear forms. In particular, by Ritz principle, the Ritz-Galerkin approximation is the best possible on V_h in the associated norm,

$$\|\tilde{u}\|_{\varepsilon,1}^{(g_\varepsilon)} = \{\int_\Omega [\varepsilon|\nabla\tilde{u}|^2 + q\tilde{u}^2]e^{-g_\varepsilon}d\Omega\}^{\frac{1}{2}}.$$

(We assume that $q \geq 0$)

Unfortunately, because of the strongly decaying exponential function, the discretization errors may be expected to be much larger away from the inflow boundary elements than they will be there.

If $\underline{v}$ is not a potential vector field we may use the method in [1], p. 337, in order to get a bilinear form which is approximately symmetrized.

Note that the basis functions in a finite element space V_h have local support. Hence we only have to evaluate the integrals with an exponential weightfunction locally. They can be accurately evaluated by use of Gaussian quadrature for exponential weight functions.

Since e^{-g_ε} varies strongly over Ω, we must scale every row of the equations we get from (3.8). Unfortunately, this means that we will transform the symmetric algebraic equations with matrix A, say, into a strongly nonsymmetric one, of the form DA. Alternatively we may perform a symmetric scaling on the form $D^{\frac{1}{2}} A D^{\frac{1}{2}}$, but then the right hand side (and the transformed solution) will not be properly scaled.

One sees easily (see for instance [1], p. 339-340, and [3], that the above method is a generalization of Il'ins method for one-dimensional problems (see [12]) to multidimensional ones. Related methods are discussed in [7] and [9]. In [2] it has been found that a difference form of Il'ins scheme (i.e. the original in [12]) gives pour convergence ($O(h)$), when $\varepsilon << h$, but that a finite element formulation as above and a central difference scheme both give full order ($O(h^2)$) when the solution is smooth.

4. The modified streamlined diffusion method.

We shall present a modified form of a streamlined diffusion method than that originally presented by Hughes and Brooks [11] and later discussed in [13], [18] and [10].

We shall use the notation $u_{\underline{v}} = \underline{v}\cdot\nabla u$ for the directional derivative along the streamlines. Consider at first the reduced equation

(4.1) $\quad L_0 u = u_{\underline{v}} + qu = f,\ x \in \Omega,\ u = 0$ on Γ_-.

We imbed problem (4.1) into a differential equation problem with an operator M_0 of second order, defined by

(4.2) $\quad M_0 u \equiv -\delta\nabla\cdot(\underline{v}\, L_0 u) + L_0 u = -\delta\underline{\nabla}\cdot(\underline{v}f) + f,\ x \in \Omega.$

Here $\delta > 0$ is a parameter. The boundary conditions are $u = 0$ on Γ_- and $L_0 u = f$ on Γ_+. It follows readily that the solution of (4.2) is unique and identical to the solution of (4.1).
From (4.2) follows,

$$M_0 u = -\delta\, u_{\underline{vv}} + (1-\delta\underline{\nabla}\cdot\underline{v})L_0 u - \delta\underline{v}\cdot\underline{\nabla}(qu).$$

We see that the essential new feature of M_0 as compared to L_0, is that $M_0 u$ has a diffusion term, $-\delta u_{\underline{vv}}$ acting along the streamlines.

Consider now

(4.3) $\quad L_\varepsilon u = \underline{\nabla}\cdot(-\varepsilon\underline{\nabla}u) + \underline{v}\cdot\underline{\nabla}u + qu = f,\ x \in \Omega,$

$u = 0$ on Γ_-, $\underline{\nabla u}\cdot\underline{\hat{n}} = 0$ on $\Gamma_0 \cup \Gamma_+$, which we similarly imbed into

(4.4) $\quad M_\varepsilon u \equiv -\delta\underline{\nabla}\cdot(\underline{v}L_\varepsilon u) + L_\varepsilon u = -\delta\underline{\nabla}\cdot(\underline{v}f) + f,\ x \in \Omega,$

with boundary conditions $u = 0$ on Γ_-, $\underline{\nabla u}\cdot\underline{\hat{n}} = 0$ on $\Gamma_0 \cup \Gamma_+$ and $L_\varepsilon u = f$ on Γ_+. Again, the solution of (4.4) is identical to that of (4.3). Assuming that $u \in H^2(\Omega)$, the variational formulation of (4.4) is

(4.5) $\quad a_{\delta,\varepsilon}(u,\tilde{u}) = \int_\Omega M_\varepsilon u\, \tilde{u}\, d\Omega =$

$$= \int_\Omega [\delta\, L_\varepsilon u\, \tilde{u}_{\underline{v}} + \varepsilon\underline{\nabla u}\cdot\underline{\nabla\tilde{u}} + (u_{\underline{v}}+qu)\tilde{u}]d\Omega$$

$$= \int_\Omega f(\tilde{u}+\delta\tilde{u}_{\underline{v}})d\Omega\ \forall\ \tilde{u} \in \mathring{H}^1(\Omega)\ \text{(trace of } \tilde{u} = 0 \text{ on } \Gamma_-).$$

Note that both $\underline{v}\cdot\underline{\hat{n}} = 0$ on $\Gamma_0 \cup \Gamma_+$ and $L_\varepsilon u = f$ on Γ_+ are natural boundary conditions for this variational formulation. We see that, for the above boundary conditions, (4.5) is identical to the Petrov-Galerkin variational formulation in [18].

Let $T = \{\tau\}$ be a triangulation of Ω and $V_h \subset \mathring{H}^1(\Omega)$ a finite dimensional subspace consisting of piecewise polynomials of degree k on T. For the Galerkin formulation of (4.5) we have to observe that the leading term doesn't exist (unless $V_h \subset H^2(\Omega)$). Hence that term has to be kept as a sum of integrals over each individual element (see also [18] and [10]). We then get

(4.6) $$a_{\delta,\varepsilon}(u_h,\tilde{u}) = \sum_{\tau\in T} \int_\tau \delta\underline{\nabla}\cdot(-\varepsilon\underline{\nabla}u_h)\tilde{u}_{\underline{v}}\,d\tau +$$

$$\int_\Omega [\delta(u_h)_{\underline{v}}\tilde{u}_{\underline{v}} + \varepsilon\underline{\nabla}u_h\cdot\underline{\nabla}\tilde{u} + \delta q\, u_h\,\tilde{u}_{\underline{v}} + ((u_h)_{\underline{v}}+qu_h)\tilde{u}]d\Omega$$

$$= \int_\Omega f(\tilde{u}+\delta\tilde{u}_{\underline{v}})d\Omega \;\; \forall\, \tilde{u} \in V_h,$$

where u_h is the associated (streamlined diffusion) Galerkin approximation of u.

In the following, for notational simplicity, we shall assume that ε is a scalar and q is constant. Note then, that if V_h consists of piecewise linear finite elements, the first term in (4.6) vanishes.

Lemma 4.1. Assume that $q - \frac{1}{2}(1+\delta q)\underline{\nabla}\cdot\underline{v} \geq 0$, $\underline{x} \in \Omega$, that $1 + \delta q \geq \frac{1}{2}$ and that $\varepsilon\delta \leq c_0 h^2$, where c_0 is a sufficiently small positive number. Then $a_{\delta,\varepsilon}(\cdot,\cdot)$ is coercive (positive definite) on $V_h \times V_h$.

Proof. By a Green's formula we have

$$a_{\delta,\varepsilon}(\tilde{u},\tilde{u}) = \sum_{\tau\in T} \int_\Omega -\delta\varepsilon\underline{\nabla}^2\tilde{u}\,\tilde{u}_{\underline{v}}\,d\Omega + \int_\Omega [\delta(\tilde{u}_{\underline{v}})^2 + \varepsilon|\underline{\nabla}\tilde{u}|^2 +$$

$$(q - \frac{1}{2}(1+\delta q)\underline{\nabla}\cdot\underline{v})\tilde{u}^2]d\Omega + \oint_{\Gamma_+} (1+\delta q)\underline{v}\cdot\hat{\underline{n}}\,\tilde{u}^2 d\Gamma.$$

We now use the inverse inequality (valid on V_h only),

$\|\tilde{u}\|_{2,\tau} \leq C_0 h^{-1}\|\tilde{u}\|_{1,\tau} \;\; \forall\, \tilde{u} \in V_h$, where $\|\tilde{u}\|_{s,\tau}$ is the Sobolev norm of order s on τ. Then by standard inequalities we get

$$\left|\int_\tau \delta\varepsilon\underline{\nabla}^2\tilde{u}\,\tilde{u}_{\underline{v}}\,d\tau\right| \leq \frac{1}{2}\,\delta \int_\tau (\tilde{u}_{\underline{v}})^2 d\tau + \frac{1}{2}\,\delta\varepsilon^2 C_0^2 h^{-2}\|\tilde{u}\|^2_{1,\tau},$$

so with $c_0 \leq C_0^{-2}$, we get

(4.7) $$a_{\delta,\varepsilon}(\tilde{u},\tilde{u}) \geq \frac{1}{2}\{\int_\Omega[\delta(\tilde{u}_{\underline{v}})^2 + \varepsilon|\underline{\nabla}\tilde{u}|^2]d\Omega + \oint_{\Gamma_+} \underline{v}\cdot\hat{\underline{n}}\,\tilde{u}^2 d\Gamma\}^{\frac{1}{2}},$$

$\forall\, \tilde{u} \in V_h$, which proves coercivity.

Corollary 4.1. Under the assumptions of Lemma 4.1, the Galerkin variational formulation (4.6) of problem (4.4) (i.e. also of (4.3)) has a unique solution u_h.

Proof: By Lemma 4.1, the stiffness matrix associated with (4.6) is nonsingular.

Note that the conditions of Lemma 4.1 are satisfied if $q \ge 0$ and $\underline{\nabla}\cdot\underline{v} \le 0$. The case $q \equiv 0$, $\underline{\nabla}\cdot\underline{v} \equiv 0$ appears frequently in applications.

To prove a discretization error estimate we let as before u_{I_h} be the interpolant of u on V_h, $\eta = u-u_{I_h}$, the interpolation error and $\theta = u_h-u_{I_h}$.

Theorem 4.1. Let u_h be defined by (4.6). Under the assumption of Lemma 4.1, the discretization error $u-u_h$ satisfies

$$(4.8)\qquad \{\int_\Omega [\delta(u-u_h)_{\underline{v}}^2 + \varepsilon|\underline{\nabla}(u-u_h)|^2]d\Omega + \oint_{\Gamma_+} \underline{v}\cdot\underline{\hat{n}}(u-u_h)^2 d\Gamma\}^{\frac{1}{2}}$$

$$\le C[\delta^{\frac{1}{2}} h^{s_1} \|u_{\underline{v}}\|_{s_1} + (\varepsilon^{\frac{1}{2}} h^s + \delta^{-\frac{1}{2}} h^{s+1})\|u\|_{s+1}]$$

$$\le C[(\delta^{\frac{1}{2}}+\varepsilon^{\frac{1}{2}})h^s + \delta^{-\frac{1}{2}} h^{s+1}]\|u\|_{s+1}, \text{ if } 0 < s \le s_1 \le k,$$

for some constant C (independent of h, δ, ε).

Proof: Similarly to Section 3, we use the relation

$$(4.9)\qquad a_{\delta,\varepsilon}(\theta,\theta) = a_{\delta,\varepsilon}(\eta,\theta) =$$

$$\sum_{\tau\in T} \int_\tau \delta\underline{\nabla}\cdot(-\varepsilon\underline{\nabla}\eta)\theta_{\underline{v}} d\tau + \int_\Omega [\delta\eta_{\underline{v}}\theta_{\underline{v}} + \varepsilon\underline{\nabla}\eta\cdot\underline{\nabla}\theta + \delta q\eta\theta_{\underline{v}} +$$

$$+ \eta_{\underline{v}}\theta + q\eta\theta]d\Omega.$$

We consider the first term. We have

$$|\sum_\tau \int_\tau \varepsilon\delta\underline{\nabla}^2\eta\, \theta_{\underline{v}} d\tau| \le \varepsilon^2\delta \sum_\tau \|\eta\|_{2,\tau}^2 + \frac{1}{4}\delta\|\theta_{\underline{v}}\|^2$$

$$\le c_0\, \varepsilon\, h^2 \sum_\tau \|\eta\|_{2,\tau}^2 + \frac{1}{4}\delta\|\theta_{\underline{v}}\|^2,$$

where we have used $\varepsilon\delta \le c_0 h^2$. For the term $\int_\Omega \eta_{\underline{v}}\theta d\Omega$ we use at first Green's formula,

$$\int_\Omega \eta_{\underline{v}}\theta d\Omega = -\int_\Omega [\eta\theta_{\underline{v}} + \underline{\nabla}\cdot\underline{v}\eta\theta]d\Omega + \oint_{\Gamma_+} \underline{v}\cdot\underline{\hat{n}}\,\eta\theta d\Gamma$$

and the bound

$$|\int \eta\theta_{\underline{v}} d\Omega| \le \delta^{-1} \int \eta^2 d\Omega + \frac{1}{4}\delta \int \theta_{\underline{v}}^2 d\Omega.$$

By standard inequalities, it now follows from (4.9),

$$a_{\delta,\varepsilon}(\theta,\theta) \le C[\delta\|\eta_{\underline{v}}\|^2+\varepsilon h^2\sum_\tau\|\eta\|^2_{2,\tau}+\varepsilon\|\eta\|^2_1+\delta^{-1}\|\eta\|^2+\oint_{\Gamma_+}\eta^2 d\Gamma]$$

and by (4.7), a triangle inequality and

$$\sum_\tau \|\eta\|_{r,\tau} \le Ch^{s+1-r}\|u\|_{s+1}\quad,\quad 0 \le r \le s \le k,$$

we get (4.8).

Note that the regularity along the streamlines is usually higher than across the streamlines (see e.g. [4]). Hence it is natural to assume that $s_1 \ge s$.

Remark 4.1. It follows from (4.8) that for the discretization error $\|(u-u_h)_{\underline{v}}\|$ we have an error estimate of optimal order, $O(h^s)$, for any fixed value of δ, ($\delta \le 1$) if $\varepsilon \le c_0 h^2$. By a Sobolev inequality, this implies an error estimate in maximum norm (and hence also in L_2-norm) of the same (in this case nonoptimal) order $O(h^s)$.

Remark 4.2. If $q \ge 0$ and $q_0 = \min\{q - \frac{1}{2}(1+\delta q)\ \underline{\nabla}\cdot\underline{v}\} > 0$, $\underline{x} \in \Omega$ we have $a_\delta(\theta,\theta) \ge c \int \theta^2 d\Omega$, for some positive constant c. In that case we also get an error estimate in L_2-norm of $O(h^{s+\frac{1}{2}})$, if we choose $\delta = h$. In [18] it is proven that by use of a smooth weight function, one may derive such an estimate (for $\delta = h$) even if $q_0 = 0$. Alternatively, we may use the method in [1] with a smooth exponential weight function to transform the given problem into a new one, for which the corresponding value of $q_0 > 0$.

5. Conclusions.

The essential feature of the method presented in section 4 is to get a diffusion term without perturbing the solution. In this way we get a Galerkin approximation which converges arbitrarily rapidly to the exact solution and with an error of (almost) optimal order when $h \to 0$, if only k is large enough and the solution is sufficiently smooth. If the solution contains a travelling shock wave, along a characteristic line this will not be smeared out, because the diffusion acts only along the streamlines.

Comparing the results of Theorems 3.1 and 4.1, we see that for

$h \le \varepsilon$, there is no gain in applying the streamlined diffusion method. If $\varepsilon < h$, the L_2-error estimate is of higher order in the latter method, and at most $O(h^{\frac{1}{2}})$ higher, when $\varepsilon \le h^2$. Furthermore, in the latter method we have an error estimate $\|(u-u_h)_{\underline{v}}\|$ and hence also in maximum norm.

The only problem which persists is that, when $\varepsilon \le O(h^2)$, there may still arise oscillations across characteristic layers, because the solution is not smooth there and the mesh is too coarse to be able to resolve these layers. To some extent it may be resolved by a say, exponentially, graded mesh (cf. Section 1), but this can only be applied if the location of the layer is known. For interior layers this is usually not the case.

As a remedy to this we advocate to use a defect-correction method of the following type.

Consider a sequence of iterates $u_h^{(\ell)}$, $\ell = 0,1,\ldots$ defined by

$$(5.1)\qquad a_{\delta,\tilde{\varepsilon}}(e_h^{(\ell)},\tilde{u}) = -a_{\delta,\varepsilon}(u_h^{(\ell)},\tilde{u}) + (f,\tilde{u}+\delta\tilde{u}_{\underline{v}})\ \forall\ \tilde{u} \in \mathring{H}^1(\Omega),$$

$$u_h^{(\ell+1)} = u_h^{(\ell)} + e_h^{(\ell)}, \quad \ell = 0,1,\ldots$$

Here we may let $u_h^{(0)}$ be derived from a standard artificial diffusion method, such as the Galerkin solution of

$$L_{\varepsilon_0} u^{(0)} = f, \quad \text{where } \varepsilon_0 = \varepsilon + h.$$

The corrections $e_h^{(\ell)}$ in (5.1) are calculated from a bilinear form $a_{\delta,\tilde{\varepsilon}}(.,.)$ where we have applied a variable artificial diffusion (in addition to the streamlined diffusion) with

$$\tilde{\varepsilon} = \varepsilon + ch^2|L_\varepsilon u_h^{(\ell)}-f|_\tau, \quad \underline{x} \in \tau,\ \tau \in T.$$

Here $|L_\varepsilon u_h^{(\ell)}-f|_\tau$ indicates some average value of $L_\varepsilon u_h^{(\ell)}-f$ on element τ, and c is a positive parameter. In this way we apply an artificial diffusion in a particular element essentially only if the approximation $u_h^{(\ell)}$ is not already a reasonably good approximation. Hence, essentially it is only applied in the layer elements and the right amount of artificial diffusion is automatically handled by the method.

The arguments for this method is still only on the heuristic side. For a recent discussion of a defect-correction method with L_{ε_0} as corrector operator, see [4].

Note that in defect-correction methods for singularly perturbed problems (the type of problem we are dealing with) the number of iterations should be very small, one or two suffices frequently. If one iterates too far, the undesired oscillations in the approximate solution may appear when the defect is getting smaller.

References

1. O. Axelsson, Stability and error estimates of Galerkin finite element approximations for convection-diffusion equations, IMA J. Num. Anal. 1 (1981), 329-345.

2. O. Axelsson, On the numerical solutions of convection-diffusion equations, Proceedings, XV Semester Computational Mathematics 1980, Banach Center Publications, Warsaw, to appear.

3. O. Axelsson, On the numerical solution of convection-diffusion equations, Lecture Note EPFL, Lausanne, March 1980.

4. O. Axelsson and W. Layton, Defect correction methods for convection dominated convection-diffusion problems, Report 8335, Department of Mathematics, Catholic University, Nijmegen.

5. O. Axelsson and G. Carey, On the stability of convection-diffusion problems, in preparation.

6. O. Axelsson and R. Masenge, Numerical treatment of boundary layers in two-point boundary value problems, in preparation.

7. J.W. Barrett and K.W. Morton, Optimal finite element solutions to diffusion-convection problems in one dimension, Int. J. Num. Math. Engng., 15 (1980), 1457-1474.

8. R.E. Ewing and M.F. Wheeler, Computational aspects of mixed finite element methods, Numerical Methods for Scientific Computing (R.S. Stepleman, ed.) North Holland, 1983.

9. E.C. Gartland, Jr., Discrete weighted mean approximation of a model convection-diffusion equation, SIAM J. Sci. Stat. Comput. 3, 460-472.

10. P.W. Hemker, Numerical aspects of singular perturbation problems, Report NW 133/82 Mathematical Centre, Amsterdam, 1982.

11. T.J. Hughes and A. Brooks, A multidimensional upwind scheme with no crosswind diffusion, in AMD vol. 34, Finite element methods for convection dominated flows, T.J. Hughes (ed.), ASME, New York, 1979.

12. A.M. Il'in, Differencing scheme for a differential equation with a

small parameter affecting the highest derivative, Math. Notes, 6 (1969), 596-602.

13. C. Johnson and U. Nävert, An analysis of some finite element methods for advection-diffusion problems, in O. Axelsson, L.S. Frank and A. Van der Sluis, (eds.), Analytical and numerical approaches to asymptotic problems in Analysis, North Holland, 1981.

14. R.B. Kellogg and H. Han, The finite element method for a singular perturbation problem using enriched subspaces, Technical note BN-978, Institute for physical science and Technology, University of Maryland, 1981.

15. F. Kikuchi, Discrete maximum principle and artificial viscosity in finite element approximations to convective diffusion equations. Report No. 550, Institute of Space and Aeronautical Science, University of Tokyo, 1977.

16. R.E. O'Malley, Jr., Topics in singular perturbations, Advances in Mathematics, 2 (1967/68), pp. 365-470.

17. A.R. Mitchell and D.F. Griffiths, Upwinding by Petrov-Galerkin methods in convection-diffusion problems, J. Comp. and Appl. Maths. 6 (1980), 219-228.

18. U. Nävert, A finite element method for convection-diffusion problems, Thesis, Chalmers University of Technology, Gothenburg, Sweden, 1982.

19. G.D. Raithby, Skew upstream differencing schemes for problems involving fluid flow. Comp. Math. Appl. Mech. Engng. 9 (1976), 153-164.

THE GROUP OF MOTIONS IN THE PLANE AND SEPARATION OF VARIABLES IN CYLINDRICAL COORDINATES

ROBERT BURRIDGE*

Abstract. In this paper a scheme is developed for handling tensor partial differential equations which are invariant under the group M(2) of motions of the Euclidean plane in relation to cylindrical polar coordinates. The work closely resembles that presented in Burridge [1] which concerned the rotation group and the treatment of tensor partial differential equations with spherical symmetry in spherical polar coordinates.

This technique involves the definition of scalar-valued functions on the group corresponding to tensor-valued functions in the plane and the expansion of these in terms of the entries in irreducible matrix representations of the group with coefficients depending on out-of-the-plane coordinates.

We find the effect of covariant differentiation on these out-of-the-plane coefficients, the functions on the group are fully determined by the tensor indices and two further parameters which remain invariant. Any given M(2)-invariant partial differential operator is then regarded as a sequence of covariant differentions and contractions with M(2)-invariant tensors.

We illustrate the method by separating variables in the elastodynamic equations for a stratified medium and deriving the system of ordinary differential equations obeyed by the z-dependent coefficients, (x,y) or (r,ϕ) being coordinates in the Euclidean plane whose motions leave the system invariant.

1. Introduction. We develop a scheme for dealing with tensor partial differential equations which are invariant under the group M(2) of Euclidean motions of the plane with reference to cylindrical coordinates.

---

*Courant Institute of Mathematical Sciences, New York University, NY 10012. This work was supported by the National Science Foundation Earth Sciences Division under Grant No. EAR-80-07816 and the Office of Naval Research under Contract No. N00014-83-K-0144.

In the special case where the dependent variables are scalars it is well known that it is useful to express a solution of such a system, say M, as a superposition of products of Bessel functions and trigonometric functions (or complex exponentials):

$$(1.1)\qquad M(r,\phi,z) = \sum_{n=-\infty}^{\infty} \int_0^{\infty} M_n(k,z)\, J_n(kr)\, e^{in\phi} k\, dk\ .$$

Here M(2) acts in the (x,y) plane, where x,y,z are cartesian coordinates; r, ϕ, z are cylindrical coordinates.

Substitution of M into an M(2)-invariant differential equation, say

$$(1.2)\qquad \Delta M + K^2(z)M = 0\ ,$$

leads to a decoupled set of ordinary differential equations in z for the $M_n(k,z)$:

$$(1.3)\qquad \frac{d^2}{dz^2} M_n(k,z) + (K^2(z)-k^2)M_n(k,z) = 0\ ,\ -\infty < n < \infty\ ,\quad 0 \leq k < \infty\ ,$$

in which the $M_n(k,z)$ for different values of k,n are decoupled.

When the dependent variables are vectors or tensors, expansion of the components in the set of functions $\{J_n(kr)e^{in\phi}\}$ is not convenient. Not only are the expressions for the differential operators combersome in terms of cylindrical coordinates but the functions with different k,n are coupled together.

We shall show, however, that if a tensor component in cylindrical coordinates is expanded in terms of the

$$(1.4)\qquad J_{n-m}(kr)e^{in\phi}\ ,$$

where m is determined by the tensor index, then the equations with different values of k,n again decouple.

The tensor components which admit this procedure are not the r and ϕ components themelves, but constant complex linear combinations of them called canonical components. Thus for a vector (M_r,M_ϕ,M_z) we would use

$$(1.5)\qquad \begin{aligned} M^+ &= M_r - iM_\phi \\ M^0 &= M_z \\ M^- &= -M_r - iM_\phi\ . \end{aligned}$$

Then we may usefully write

$$(1.6) \qquad M^{\lambda} = \sum_{n=-\infty}^{\infty} \int_0^{\infty} M_n^{\lambda}(k,z)\, J_{n-\lambda}(kr)\, k\, dk\, e^{in\phi}$$

where $\lambda = +,0,-$ and $+,-$ are abbreviations for $1,-1$. The functions

$$(1.7) \qquad T^k_{mn}(r,\phi,\psi) = e^{im\psi}\, J_{n-m}(kr)\, e^{in\phi}$$

arise as matrix entries in the unitary representations of the group M(2). The different representations are labeled by k; m,n are the matrix subscripts. The variables r,ϕ,ψ are analogous to the Euler angle θ,ϕ,ψ parameterizing the rotation group. Indeed, following Burridge [1], Gelfand & Shapiro [2] who performed a similar analysis for spherical systems, we define a (scalar) function on the group M(2) for each canonical component in space (see (1.4)). It is found that the tensor index $\lambda\mu\nu\ldots$ of the field $M^{\lambda\mu\nu\ldots}$ in question determines the dependency of these fields on the angle ψ; in fact $M^{\lambda\mu\nu\ldots}$ varies as $e^{im\psi}$, where $m = \lambda+\mu+\nu\ldots$. Finally, specializing to $\psi = 0$ has the effect of referring the tensor components to the coordinates r,ϕ,z in space rather than the group.

Any M(2)-invariant partial differential operator on space may be regarded as a sequence of covariant differentiations, contractions with constant M(2)-invariant tensors, and multiplication by functions of z or a sum of such operators. We find it convenient to express covariant differentiation in terms of the infinitesimal motions of M(2). We then use the structure of M(2) as elucidated by Vilenkin [3] to obtain the effect of covariant differentiation on the coefficients of the functions (1.3). Actually z appears only as a parameter in most of the analysis and so in Sections 2 through 9 we consider only the dependence on r,ϕ. It is straightforward to add the effects of z-dependence later in the applications.

In section 2 we introduce the group M(2) with its most natural parameterization. The canonical basis, relative to which the components are indexed by $\lambda\mu\ldots$; $\lambda,\mu = \pm$, is introduced in section 3. In section 4 we explain how M(2) acts on tensor fields defined on E_2 the Euclidean plane. We then define the corresponding canonical scalar fields on M(2) and how M(2) acts on them. In section 5 we relate polar coordinates in E_2 to the 'Euler angle' parameterization of M(2). In section 6 we explain how M(2) acts on the fields defined on M(2) and the effect of covariant differentiation. In section 7, following Vilenkin [3], we deduce the form (1.7) of the matrix entries T^k_{mn}. In section 8 we examine the expansion of the canonical scalar fields in the form

$$(1.8)\qquad \overline{M}(g) = \sum_n \int M_n^{\lambda\mu\cdots}(k)\, T_{mn}^k(r,\phi,\psi) k\, dk, \qquad m = \lambda + \mu + \dots .$$

The effect of covariant differentiation of M on the coefficients $M_n^{\lambda\mu\cdots}(k)$ in this expansion is examined in section 9.

In section 10 an illustrative example drawn from elastodynamics is set up.

2. The Group of Motions of the Euclidean Plane. Let $(x,y) = (x_1,x_2)$ be cartesian coordinates in the Euclidean plane E_2. A motion of E_2 is a mapping of E_2 onto itself preserving the Euclidean distances between points. Following Vilenkin [2] let us denote by M(2) the group of all such motions which preserve orientation. Here composition is the group operation. As discussed in [2] an element of M(2) may be parameterized by an angle α $(0 \leq \alpha < 2\pi)$ and a 2-vector $\underline{a} = (a,b) = (a_1,a_2)$. Let $g(a,b,\alpha)$ be the group element corresponding to $\alpha,\underline{a}$. Then $g(a,b,\alpha)$ is interpreted as that motion of E_2 which rotates the plane about the origin counterclockwise through angle α and then translates the origin to the point $\underline{a}$ by means of a rigid parallel translation. If

$$(2.1)\qquad \underline{x}' = g(a,b,\alpha)\underline{x}$$

then

$$(2.2)\qquad \begin{aligned} x' &= x\cos\alpha - y\sin\alpha + a \\ y' &= x\sin\alpha + y\cos\alpha + b\ . \end{aligned}$$

If we further denote by g^α the rotation through α about the origin we have

$$(2.3)\qquad \underline{x}' = g^\alpha \underline{x} + \underline{a}$$

and we may identify g^α with the matrix

$$(2.4)\qquad g^\alpha = \begin{pmatrix} \cos\alpha & -\sin\alpha \\ \sin\alpha & \cos\alpha \end{pmatrix}$$

in the usual way.

Let us note that $g \in M(2)$ induces a linear representation on $L^2(E_2)$ (the regular representation) as follows: For $g \in M(2)$, $f \in L^2(E_2)$

$$(2.5)\qquad [T(g)f](\underline{x}) = f(g^{-1}\underline{x})\ .$$

For tensor-valued functions on E_2 a more complicated analogous representation may be defined: Let $M_{i_1\dots i_p}(\underline{x})$ be a tensor field on E_2 of rank p then we define

(2.6) $$[T(g)M]_{i_1 \cdots i_p}(x) = g^{\alpha}_{i_1 j_1} \cdots g^{\alpha}_{i_p j_p} M_{j_1 \cdots j_p}(g^{-1}\underline{x})$$

where g is understood to be $g(a,b,\alpha)$.

We shall need also the infinitesimal motions

(2.7) $$A_i = \left[\frac{\partial g(a_1,a_2,\alpha)}{\partial a_i} \right]_{\substack{\underline{a}=0 \\ \alpha=0}} \qquad i = 1,2 .$$

For completeness we also define

(2.8) $$A_3 = \left[\frac{\partial g(a_1,a_2,\alpha)}{\partial \alpha} \right]_{\substack{\underline{a}=0 \\ \alpha=0}} .$$

3. <u>The Canonical Basis.</u> It will be convenient to work, not in the usual basis

(3.1) $$e_1 = \begin{pmatrix} 1 \\ 0 \end{pmatrix} , \quad e_2 = \begin{pmatrix} 0 \\ 1 \end{pmatrix}$$

but in the complex basis

(3.2) $$f_+ = \frac{1}{2} \begin{pmatrix} 1 \\ i \end{pmatrix} , \quad f_- = \frac{1}{2} \begin{pmatrix} -1 \\ i \end{pmatrix} .$$

We note that f_λ is an eigenvector of A_3 and of g^α, $\lambda = \pm$. Here, as before, $\pm$ is an abbreviation of ± 1. Specifically

(3.3) $$A_3 f_\lambda = - i\lambda f_\lambda , \quad g^\alpha f_\lambda = e^{-i\lambda\alpha} f_\lambda .$$

The bases e and f are related by

(3.4) $$f_\lambda = \frac{1}{2} \overline{C}_{\lambda i} e_i , \quad e_i = C_{\lambda i} f_\lambda$$

where

(3.5) $$C = (C_{\lambda i}) = \begin{pmatrix} 1 & -i \\ -1 & -i \end{pmatrix} .$$

Note that

$$C^{-1} = \frac{1}{2} C^{\top} = \frac{1}{2} \begin{pmatrix} 1 & -1 \\ i & i \end{pmatrix} . \tag{3.6}$$

Here and subsequently Greek subscripts $\lambda, \mu, \ldots$ range over $+,-$ or sometimes later $+,0,-$; Roman subscripts range over 1,2 or sometimes later 1,2,3; the Einstein summation convention applies. If

$$x_i e_i = x^{\lambda} f_{\lambda} \tag{3.7}$$

then

$$x^{\lambda} = C_{\lambda i} x_i . \tag{3.8}$$

We shall be concerned with the corresponding components of tensors:

Let $M_{i_1 \ldots i_p}$ be a tensor. Then we define

$$M^{\lambda_1 \ldots \lambda_p} = C_{\lambda_1 i_1} \cdots C_{\lambda_p i_p} M_{i_1 \ldots i_p} . \tag{3.9}$$

We shall refer to these components of M as the contravariant canonical components. We also define the covariant canonical components.

$$M_{\lambda_1 \ldots \lambda_p} = C^{-1}_{i_1 \lambda_1} \cdots C^{-1}_{i_p \lambda_p} M_{i_1 \ldots i_p} . \tag{3.10}$$

We shall not need the covariant components so much for general tensor fields as for certain constant isotropic tensors which will be contracted with general fields. In particular the following tensor components will be useful.

Let

$$M_{i_1 i_2} = \delta_{i_1 i_2} , \quad \text{the Kronecker symbol.} \tag{3.11}$$

then we define by (3.10)

$$\begin{aligned} e_{\lambda\mu} \equiv M_{\lambda\mu} &= C^{-1}_{i\lambda} C^{-1}_{j\mu} \delta_{ij} \\ &= C^{-1}_{i\lambda} C^{-1}_{i\mu} \\ &= \frac{1}{4} (\overline{C}\, C^{\top})_{\lambda\mu} \\ &= \frac{1}{2} \begin{pmatrix} 0 & -1 \\ -1 & 0 \end{pmatrix} . \end{aligned} \tag{3.12}$$

Similarly

$$(3.13)\qquad e^{\lambda\mu} \equiv C_{\lambda i}C_{\mu j}\delta_{ij} = 2\begin{pmatrix} 0 & -1 \\ -1 & 0 \end{pmatrix} .$$

Thus from (3.12), (3.13) we find

$$(3.14)\qquad 2e_{\lambda\mu} = \frac{1}{2}e^{\lambda\mu} = \begin{cases} -1, & \lambda+\mu = 0 , \\ 0, & \lambda+\mu \neq 0 . \end{cases}$$

As a simple example of how these formula are used, let us express the trace of a second-rank tensor in canonical components

$$(3.15)\qquad M_{ii} = e_{\lambda\mu}\, M^{\lambda\mu} = -\frac{1}{2}(M^{-+} + M^{+-})$$

where we have used (3.14).

4. The Motion of Tensor Fields and Canonical Scalar Fields on M(2).

Let $M^{\lambda_1\ldots\lambda_p}(x)$ be the canonical contravariant components of a tensor field. Then combining (2.6) with (3.9) we find that

$$(4.1)\qquad [T(g)M]^{\lambda_1\ldots\lambda_p}(\underline{x}) = g^{\alpha}_{\lambda_1\mu_1}\cdots g^{\alpha}_{\lambda_p\mu_p}\, M^{\mu_1\ldots\mu_p}(g^{-1}\underline{x})$$

$$= \exp\left[-i\alpha\sum_{i=1}^{p}\lambda_i\right]M^{\lambda_1\ldots\lambda_p}(g^{-1}\underline{x})$$

since, by (3.3),

$$(4.2)\qquad g^{\alpha}_{\lambda\mu} = e^{-i\lambda\alpha}\,\delta_{\lambda\mu}\quad \text{(no summation)}.$$

Next, for each component of a tensor field M on E_2, we define a function on M(2) as follows

$$(4.3)\qquad \overrightarrow{M}^{\lambda_1\ldots\lambda_p}(g) = (T(g)M)^{\lambda_1\ldots\lambda_p}(0)$$

$$= \exp\left[-i\alpha\sum_{i=1}^{p}\lambda_i\right] M^{\lambda_1\ldots\lambda_p}(g^{-1}0)$$

where 0 is the origin $\underline{x} = 0$ of E_2.

5. Polar coordinates and the 'Euler-angle' parametrization of M(2). Let r, ϕ be polar coordinates in E_2. Let $g(r,\phi,\psi)$ represent a rotation through angle $-\phi$ about the origin followed by a translation in the negative x direction through a distance r (this brings the point r,ϕ to the origin) followed by a rotation through an angle $-\psi$ about the origin.

In the notation of 2.3,

$$(5.1) \qquad \alpha = -\phi - \psi \, , \quad a = -r \cos \psi \, , \quad b = r \sin \psi \, .$$

A more geometric interpretation of $g(r,\phi,\psi)$ can be given in terms of Figure 1. Thus $g^{-1}0$, which appears in (4.3), is the point r,ϕ and is independent of the angle ψ. When $\psi = 0$ the basis e_1', e_2' is the usual basis in polar coordinates (r,ϕ).

On defining the local canonical basis

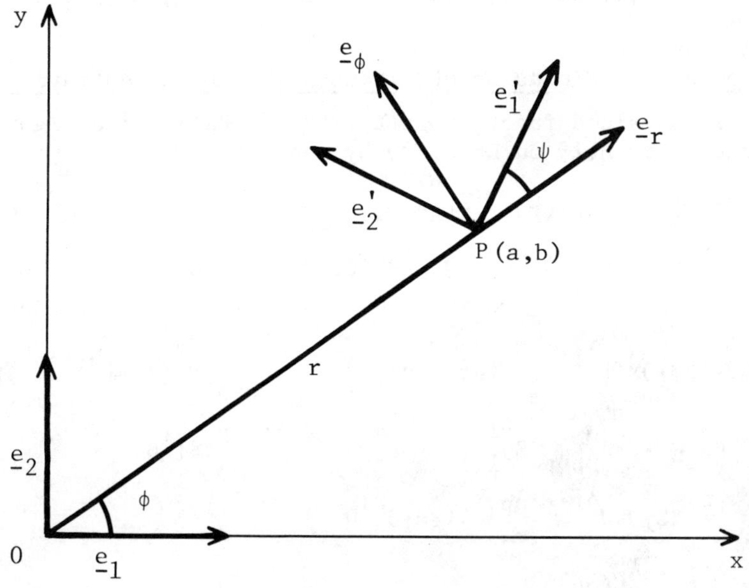

Figure 1. $g(r,\phi,\psi)$ has the effect of moving P(a,b) to 0 and the vectors e_1', e_2' into the vectors e_1, e_2.

(5.2) $$f'_\lambda = C^{-1}_{i\lambda}\, e'_i$$

where the e'_i are defined in Figure 1 we find that $\overline{M}^{\lambda_1\cdots\lambda_p}(g(r,\phi,\psi))$ may be interpreted as the canonical components of M at the point (r,ϕ) in the basis f′.

Putting together (4.3), (5.1), and the fact that $g^{-1}0$ does not depend on ψ, we see that

(5.3) $$\overline{M}^{\lambda_1\cdots\lambda_p}(g(r,\phi,\psi)) = \exp\Big[i\psi \sum_1^p \lambda_i\Big]\, M^{\lambda_1\cdots\lambda_p}(r,\phi)$$

where $M^{\lambda_1\cdots\lambda_p}(r,\phi)$ is independent of ψ. In fact, setting $\psi = 0$, we find

(5.4) $$\overline{M}^{\lambda_1\cdots\lambda_p}(g(r,\phi,0)) = M^{\lambda_1\cdots\lambda_p}(r,\phi) ,$$

where $M^{\lambda_1\cdots\lambda_p}(r,\phi)$ are components of M in the basis f'_+, f'_- related to the usual basis e_r, e_ϕ at (r,ϕ) in polar coordinates, by (3.4).

6. <u>Motion of the Fields on M(2); Covariant Differentiation.</u> Let $\overline{F}(g)$ be a scalar valued function in $L^2(M(2))$. Define the right regular representation T of M(2) on $L^2(M(2))$ by means of

(6.1) $$[\tilde{T}(g_1)\, \overline{F}](g) = \overline{F}(gg_1).$$

Two applications of (4.1), (4.3) and (6.1) lead to

(6.2) $$\{T(g)[T(g_1)M]\}^{\lambda_1\cdots\lambda_p}(0) = g^{\alpha}_{\lambda_1\mu_1}\cdots g^{\alpha}_{\lambda_p\mu_p}[T(g_1)M]^{\mu_1\cdots\mu_p}(g^{-1}0)$$

$$= g^{\alpha}_{\lambda_1\mu_1}\cdots g^{\alpha}_{\lambda_p\mu_p}\, g^{\alpha_1}_{\mu_1\nu_1}\cdots g^{\alpha_1}_{\mu_p\nu_p} M^{\nu_1\cdots\nu_p}(g_1^{-1}g^{-1}0)$$

$$= \big[T(gg_1)\, M^{\lambda_1\cdots\lambda_p}\big]\,((gg_1)^{-1}0)$$

$$= \overline{M}^{\lambda_1\cdots\lambda_p}(gg_1)$$

$$= \big[\tilde{T}(g_1)\overline{M}^{\lambda_1\cdots\lambda_p}\big](g) .$$

In particular, in the representations $\tilde{T}$ and T we may consider the infinitesimal motions $\tilde{A}_1$, $\tilde{A}_2$, A_1, A_2, and the related

(6.3) $$\tilde{A}_\lambda = C_{\lambda i}\tilde{A}_i \ , \quad A_\lambda = C_{\lambda i}A_i \ .$$

Letting $\tilde{T}(g_1)$ become $\tilde{A}_\lambda$ in (6.2) we get

(6.4) $$\lfloor \tilde{A}_\lambda \overrightarrow{M}^{\lambda_1 \ldots \lambda_p} \rfloor (g) = [T(g)(A_\lambda M)]^{\lambda_1 \ldots \lambda_p}(0) \ .$$

But using (2.7) we see that $A_i M$ is given by

(6.5) $$A_i M_{i_1 \ldots i_p} = - \frac{\partial M_{i_1 \ldots i_p}}{\partial x_i} = - M_{i_1 \ldots i_p, i} \ ,$$

minus2 the covariant derivative of M, s2o that

(6.6) $$M^{\lambda_1 \ldots \lambda_p, \lambda}(\underline{x}) = - A_\lambda M^{\lambda_1 \ldots \lambda_p}(\underline{x}) \ .$$

Thus if we lift a tensor field $M_{i_1 \ldots i_p}(\underline{x})$ and its covariant derivative $M_{i_p \ldots i_p, i}(\underline{x})$ to the group as $\overrightarrow{M}^{\lambda_1 \ldots \lambda_p}(g)$ and $\overrightarrow{M}^{\lambda_1 \ldots \lambda_p, \lambda}(g)$ then these are related by

(6.7) $$\overrightarrow{M}^{\lambda_1 \ldots \lambda_p, \lambda}(g) = - g^{\alpha}_{\lambda\mu}\tilde{A}_\mu \overrightarrow{M}^{\lambda_1 \ldots \lambda_p}(g) \ .$$

7. <u>The Matrix Entries of Irreducible Representations of</u> M(2). Here we follow Vilenkin [2, Chapter IV, Sec. 3] but make some minor changes which make our results conform more closely to the usual notation used for Hankel transforms. For each k we construct an irreducible unitary representation T^k of M(2) acting on the space $L^2(S_1)$ of square-integrable functions on the circle. Let $f(\theta)$, $0 \leq \theta < 2\pi$, be such a function and $g(a,b,\alpha) \in M(2)$ then T^k is defined by

(7.1) $$[T^k(g)f](\theta) = e^{ik(a \cos \theta + b \sin \theta)} f(\theta - \alpha) \ .$$

On using the parametrization of section 5 for g, especially equation (5.1), we rewrite (7.1) as

(7.2) $$[T^k(g)f](\theta) = e^{-ikr \cos(\theta+\psi)} f(\theta+\phi+\psi) \ .$$

In $L^2(S_1)$ we take the basis

(7.3) $$f_n(\theta) = i^n e^{in\theta} \ .$$

Then in this basis the m,n matrix entry of $T^k(g)$ is given by

$$T^k(g)f_n = \sum_m T^k_{mn}(g)\ f_m \tag{7.4}$$

$$T^k_{mn}(g) = \langle T^k(g)f_n, f_m\rangle ,$$

where $\langle\cdot,\cdot\rangle$ stands for the usual inner product in $L^2(S_1)$,

$$\langle f,g\rangle = \frac{1}{2\pi}\int_0^{2\pi} f(\theta)\ \overline{g(\theta)}\ d\theta . \tag{7.5}$$

Using (7.2), (7.3), (7.4), (7.5), we obtain

$$\begin{aligned} T^k_{mn}(g) &= \frac{1}{2\pi}\int_0^{2\pi} e^{-ikr\cos(\theta+\psi)}\ i^n\ e^{in(\theta+\phi+\psi)}\ i^{-m}\ e^{-im\theta}\ d\theta \\ &= \frac{i^{n-m}}{2\pi}\int_0^{2\pi} e^{-ikr\cos\theta}\ e^{in(\theta+\phi)}\ e^{-im(\theta-\psi)}\ d\theta \\ &= \frac{1}{2\pi}\int_0^{2\pi} e^{-ikr\cos\theta}\ e^{i(n-m)(\theta+\pi/2)}\ d\theta\ e^{im\psi+in\phi} \\ &= \frac{1}{2\pi}\int_0^{2\pi} e^{-ikr\sin\theta + i(n-m)\theta}\ d\theta\ e^{im\psi+in\phi} \\ &= J_{n-m}(kr)\ e^{im\psi}\ e^{in\phi} , \end{aligned} \tag{7.6}$$

where J_ℓ is the Bessel function of order ℓ.

By applying (2.7) to (7.1) we see that in the same representation the infinitesimal motions A^k_1, A^k_2 are

$$[A^k_1 f](\theta) = ik\cos\theta\ f(\theta) \tag{7.7}$$

$$[A^k_2 f](\theta) = ik\sin\theta\ f(\theta) .$$

Thus

$$A^k_\lambda f = C_{\lambda i}A^k_i f = i\lambda k e^{-i\lambda\theta} f. \tag{7.8}$$

In particular

$$(7.9)\qquad A^k_\lambda f_n = i\lambda k\, e^{-i\lambda\theta}\, i^n\, e^{in\theta} = -\, kf_{n-\lambda}$$

since

$$i\lambda = -\, i^{-\lambda} .$$

8. <u>The Expansion of M in Terms of the T^k_{mn}.</u> Now any function in $L^2(M(2))$ can be presented in the form

$$(8.1)\qquad \overline{M}(g) = \sum_{m=-\infty}^{\infty} \sum_{n=-\infty}^{\infty} \int_0^{\infty} M_{mn}(k)\, J_{n-m}(kr)\, e^{im\psi}\, e^{in\phi}\, k\, dk .$$

See Appendix B. In particular

$$(8.2)\qquad \overline{M}^{\lambda_1\cdots\lambda_p}(g) = e^{im\psi} \sum_{n=-\infty}^{\infty} \int_0^{\infty} M_n^{\lambda_1\cdots\lambda_p}(k)\, J_{n-m}(kr)\, k\, dk\, e^{in\phi}$$

where m takes on the single value

$$(8.3)\qquad m = \sum_{\ell=1}^{p} \lambda_\ell .$$

This is the effect of the special dependence of $\overline{M}^{\lambda_1\cdots\lambda_p}(g)$ upon ψ expressed in (5.3). On setting $\psi = 0$ and referring to (5.4) we see that

$$(8.4)\qquad M^{\lambda_1\cdots\lambda_p}(r,\phi) = \sum_{n=-\infty}^{\infty} \int_0^{\infty} M_n^{\lambda_1\cdots\lambda_p}(k)\, J_{n-m}(kr)\, k\, dk\, e^{in\phi} .$$

9. <u>The Effect of Covariant Differentiation on the Coefficients.</u> Consider the function

$$(9.1)\qquad \overline{M}^{\lambda_1\cdots\lambda_p}(g) = T^k_{mn}(g) , \qquad m = \sum_{\ell=1}^{p} \lambda_\ell$$

and let us calculate $\overline{M}^{\lambda_1\cdots\lambda_p,\lambda}$ according to (6.7) and (9.1):

$$(9.2)\qquad \overline{M}^{\lambda_1\cdots\lambda_p,\lambda}(g) = -\, g^\alpha_{\lambda\mu}\tilde{A}_\mu \overline{M}^{\lambda_1\cdots\lambda_p}(g) = -\, g^\alpha_{\lambda\mu}\tilde{A}_\mu T^k_{mn}(g) .$$

Now

$$(9.3)\qquad -\sum_m g^\alpha_{\lambda\mu}[\tilde{A}_\mu T^k_{mn}(g)]f_m = -\, g^\alpha_{\lambda\mu}\, T^k(g)\, A^k_\mu\, f_n$$

$$= -\, A^k_\lambda[T^k(g)f_n] , \qquad \text{see Appendix}$$

$$= -\sum_m A^k_\lambda[T_{mn}(g)f_m] , \qquad \text{by (7.4)}$$

$$= -\sum_m T^k_{mn}(g)\, A^k_\lambda f_m$$

$$= \sum_m T^k_{mn}(g) k f_{m-\lambda}\ , \quad \text{by (7.9)}$$

$$= \sum_m k\ T^k_{m+\lambda,n}(g)\ f_m\ .$$

Thus comparing coefficients of the f_m in the extreme members of (9.3) we obtain

(9.4) $$-\ g^{\alpha}_{\lambda\mu}\tilde{A}_{\mu}T^k_{mn}(g) = k\ T^k_{m+\lambda,n}(g)\ .$$

From (9.1), (9.2), (9.3) we get

(9.5) $$\overline{M}^{\lambda_1\ldots\lambda_p,\lambda}(g) = kT^k_{m+\lambda,n}(g)\ .$$

Hence if

(9.6) $$\overline{M}^{\lambda_1\ldots\lambda_p}(g) = M_n^{\lambda_1\ldots\lambda_p}(k)T^k_{mn}(g),\ m = \Sigma\ \lambda_\ell$$

and

(9.7) $$\overline{M}^{\lambda_1\ldots\lambda_p,\lambda}(g) = M_n^{\lambda_1\ldots\lambda_p,\lambda}(k)\ T^k_{m'n}(g),\quad m' = \lambda + \Sigma\ \lambda_\ell$$

then

(9.8) $$M_n^{\lambda_1\ldots\lambda_p,\lambda}(k) = kM_n^{\lambda_1\ldots\lambda_p}(k)\ .$$

This is our main result. Notice that, once k, n, and the tensor indices λ_ℓ are specified, the function $T^k_{mn}(g)$ is fully determined and we may concentrate upon the coefficients $M^{\lambda_1\ldots\lambda_p}(k)$. The effect of covariant differentiation can be expressed directly in terms of these coefficients. Notice also that different values of k,n are decoupled in (9.8). This is a direct consquence of: (a) Schur's lemma, (b) the fact that covariant differentiation is invariant under the action of M(2), and (c) the fact that for fixed k,m the $T^k_{mn}(g)$ form bases for function spaces in which M(2) acts irreducibly through the representation T^k. Specifically

(9.9) $$T^k(g_1)T^k_{mn}(g) = T^k_{mn}(gg_1)$$

$$= \sum_p T^k_{pn}(g_1)T^k_{mp}(g)\ .$$

10. <u>Applications</u>. This work was motivated by the need for an efficient method of generating the systems of ordinary differential equations in z which arise when a tensor partial differential equation which is invariant under motions of the xy plane are separated in cylindrical polar coordinates.

A typical example is provided by the equations of elastodynamics assuming a harmonic time dependence.

Let x,y,z be Cartesian coordinates. Consider a stratified isotropic elastic medium of density $\rho(z)$ and elastic constants $\lambda(z)$, $\mu(z)$ to occupy the half space $z > 0$. The momentum and constitutive equations of such a system are

$$(10.1) \qquad i\omega\rho v_i = \tau_{ij,j}$$

$$(10.2) \qquad i\omega\tau_{ij} = \lambda v_{k,k}\delta_{ij} + \mu(v_{i,j} + v_{j,i})$$

where v_i is particle velocity and τ_{ij} is the stress tensor, and subscripts run from 1 to 3. Rewriting (10.1), (10.2) in canonical cylindrical components we have

$$(10.3) \qquad \begin{aligned} i\omega\rho v^{\alpha} &= e_{\beta\gamma}\, \tau^{\alpha\beta,\gamma} \\ i\omega\tau^{\alpha\beta} &= \lambda\, e_{\gamma\delta}\, v^{\gamma,\delta}\, e^{\alpha\beta} + \mu(v^{\alpha,\beta} + v^{\beta,\alpha}) \ . \end{aligned}$$

Set

$$(10.4) \qquad \begin{aligned} v^{\alpha} &= V^{\alpha}(z)\, J_{n-\alpha}(kr)\, e^{in\phi} \\ v^{\gamma,\delta} &= V^{\gamma,\delta}(z)\, J_{n-\gamma-\delta}(kr)\, e^{in\phi} \\ \tau^{\alpha\beta,\gamma} &= T^{\alpha\beta\ \gamma}(z)\, J_{n-\alpha-\beta-\gamma}(kr)\, e^{in\phi} \end{aligned}$$

where now $\alpha,\beta,\ldots$ range over +,0,− . Also

$$(10.5) \qquad e_{\alpha\beta} = \begin{pmatrix} 0 & 0 & -\frac{1}{2} \\ 0 & 1 & 0 \\ -\frac{1}{2} & 0 & 0 \end{pmatrix}, \qquad e^{\alpha\beta} = \begin{pmatrix} 0 & 0 & -2 \\ 0 & 1 & 0 \\ -2 & 0 & 0 \end{pmatrix} .$$

In (10.4) we have suppressed the symbols k, n on V^{α}, $T^{\alpha\beta}$. In general v^{α}, $\tau^{\alpha\beta}$ will be linear combinations of terms like (10.4) with differing values of k,n. Then by (9.8)

$$(10.6) \qquad \begin{aligned} V^{\gamma,\delta} &= kV^{\gamma} \quad \text{for } \delta = \pm \\ &= V^{\gamma}_{z} \quad \text{for } \delta = 0 \end{aligned}$$

where $\delta = 0$ denotes the z-direction and the subscript z denotes differentiation in z. We shall need

$$(10.7) \qquad \begin{aligned} e_{\gamma\delta}V^{\gamma,\delta} &= V^{0,0} - \frac{1}{2}(V^{+,-} + V^{-,+}) \\ &= V^{0}_{z} - \frac{k}{2}(V^{+} + V^{-}) \ . \end{aligned}$$

Similarly

$$(10.8) \qquad T^{\alpha\beta,\gamma} = kT^{\alpha\beta} , \quad \gamma = \pm ,$$

$$T^{\alpha\beta,0} = T^{\alpha\beta}_z .$$

$$(10.9) \qquad e_{\beta\gamma}T^{\alpha\beta,\gamma} = T^{\alpha 0}_z - \frac{1}{2}(T^{\alpha+,-} + T^{\alpha,+}) = T^{\alpha 0}_z - \frac{k}{2}(T^{\alpha+} + T^{\alpha-}) .$$

Now (10.3) lead to equations involving only the z-dependent coefficients $V^{\alpha}(z)$, $T^{\alpha\beta}(z)$:

$$(10.10a) \qquad i\omega\rho V^{\alpha} = T^{\alpha 0}_z - \frac{k}{2}(T^{\alpha+} + T^{\alpha-})$$

$$(10.10b) \qquad i\omega T^{\alpha\beta} = \lambda(V^0_z - \frac{k}{2}(V^+ - V^-))e^{\alpha\beta} + \mu(V^{\alpha,\beta} + V^{\beta,\alpha})$$

$$(10.11a) \qquad i\omega\rho V^0 = T^{00}_z - \frac{k}{2}(T^{0+} + T^{0-})$$

$$(10.11b) \qquad i\omega\rho V^+ = T^{+0}_z - \frac{k}{2}(T^{++} + T^{+-})$$

$$(10.11c) \qquad i\omega\rho V^- = T^{-0}_z - \frac{k}{2}(T^{-+} + T^{--})$$

$$(10.11d) \qquad i\omega T^{00} = \lambda[V^0_z - \frac{k}{2}(V^+ + V^-)] + 2\mu V^0_z$$

$$(10.11e) \qquad i\omega T^{0+} = \mu(V^+_z + kV^0)$$

$$(10.11f) \qquad i\omega T^{0-} = \mu(kV^0 + V^-_z)$$

$$(10.11g) \qquad i\omega T^{+-} = -2\lambda[V^0_z - \frac{k}{2}(V^+ + V^-)] + k(V^+ + V^-)$$

$$(10.11h) \qquad i\omega T^{++} = 2\mu k V^+$$

$$(10.11i) \qquad i\omega T^{--} = 2\mu k V^-$$

This is a system of nine equations satisfied by the V^{α} and $T^{\alpha\beta}$, and this completes the illustration of the method advocated in this paper. However, this system is of intrinsic interest and we continue to reduce it further.

Actually T^{++}, T^{--} are expressed algebraically in terms of V^+, V^- by (10.11h,i). An algebraic expression for T^{+-} in terms of V^+, V^-, T^{00} may be obtained by eliminating V^0_z between (10.11d,g):

$$(10.12) \qquad T^{+-} = \frac{\mu(3\lambda+2\mu)}{\lambda + 2\mu}\frac{k}{i\omega}(V^+ + V^-) - \frac{2\lambda}{\lambda+2\mu}T^{00}$$

Making use of (10.11h,i) and (10.12) enables us to eliminate T^{--}, T^{++}, T^{+-}, T^{-+} from the system (10.11a-f). The system obtained is a 6 × 6 system of ordinary differential equations for the V^{α} and the $T^{0\alpha}$. Before writing this system let us note the invariance of the system under the interchange of +,-. We now form differences and sums of equations which differ only by the interchange of +,-. Thus (10.11b,c) and (10.11e,f) yield

$$(10.13)\qquad (T^{0+} - T^{0-})_z = \frac{1}{i\omega}(\mu k^2 - \rho\omega^2)(V^+ - V^-)$$
$$(V^+ - V^-)_z = \frac{i\omega}{\mu}(T^{0+} - T^{0-}) .$$

Equations (10.13) are a 2×2 system which decouples from the rest; they govern the propagation of SH and Love waves and are analogous to the toroidal modes in spherical systems. No vertical motions (V^0) appear in this system and only the shear modulus μ (but not λ) is involved. On adding pairs of equations symmetric in +,−, the resulting equations form a system of four ordinary differential equations for V^0, $V^+ + V^-$, T^{00}, $T^{0+} + T^{0-}$. After some calculation one finds that

$$(10.14)\qquad V^0_z = \frac{\lambda k}{2(\lambda+2\mu)}(V^+ + V^-) + \frac{1}{\lambda+2\mu} i\omega T^{00}$$
$$(V^+ + V^-)_z = -2kV^0 + \frac{1}{\mu} i\omega(T^{0+} + T^{0-})$$

$$T^{00}_z = i\omega\rho V^0 + \frac{k}{2}(T^{0+} + T^{0-})$$
$$(T^{0+} + T^{0-})_z = \frac{1}{i\omega}\left\lfloor \frac{4\mu(\lambda+\mu)}{\lambda + 2\mu} k^2 - \rho\omega^2 \right\rfloor (V^+ + V^-) - \frac{2\lambda}{\lambda+2\mu} kT^{00} .$$

This system, which represents coupled vertical and horizontal motion, governs the propagation of P, SV and Rayleigh waves, which are analogous to the spheroidal modes in spherical systems.

If one is concerned solely with dispersion relations then the eigenproblem for systems (10.13) or (10.14) with appropriate boundary conditions are sufficient to give ω as a function $\omega(k)$ of k. In other situations it may be necessary to transform back to the (noncanonical) physical r,ϕ,z-components.

Acknowledgments. I wish to thank Professor Gross for the many helpful conversations we had during the summer of 1982 concerning harmonic analysis on groups and on more specific topics related to this work.

I gratefully acknowledge financial support from the National Science Foundation, Earth Sciences Division under Grant No. EAR-80-07816 and the Office of Naval Research under Contract No. N00014-83-K-0144.

Appendix A. In the string of equalities (9.3) we needed the identity

$$(A.1)\qquad g^{\alpha}_{\lambda\mu} T^k(g) A^k_\mu = A^k_\lambda T^k(g) .$$

To verify this we use (7.1) and (7.8)

$$(A.2)\qquad A^k_\lambda T^k(g) f(0) = i\lambda k e^{-i\lambda\theta} e^{ik(a\cos\theta + b\sin\theta)} f(\theta-\alpha).$$

Also

(A.3) $$A_\mu^k f(\theta) = i\mu k e^{-i\mu\theta} f(\theta)$$

(A.4) $$T^k(g)A_\mu^k f(\theta) = i\mu k e^{ik(a\cos\theta + b\sin\theta)}\, e^{-i\mu(\theta-\alpha)} f(\theta-\alpha).$$

But repeating (4.2) we have

(A.5) $$g_{\lambda\mu}^\alpha = e^{-i\lambda\alpha}\,\delta_{\lambda\mu}$$

and so

(A.6) $$g_{\lambda\mu}^\alpha T^k(g)A_\mu^k(\theta) = i\lambda k e^{ik(a\cos\theta + b\sin\theta)}\, e^{-i\lambda\theta}\, f(\theta-\alpha)\ .$$

The identity (A.1) now follows from (A.2) and (A.6).

Appendix B. Fourier analysis on M(2). We shall regard M(2) as the cartesian product $E_2 \otimes S_1$ of the Euclidean plane and the circle.

The space $L^2(E_2)$ is the continuous direct sum

(B.1) $$\int_k \sum_n \otimes\, Y_n^k$$

where $Y_n^k(r,\phi)$ is the subspace spanned by $J_n(kr)e^{in\phi}$, (r,ϕ) being polar coordinates in E_2. The space $L^2(S_1)$ is the direct sum

(B.2) $$\sum_m \otimes\, Z_m$$

where Z_m is the space spanned by $e^{im\alpha}$, $\alpha \in [0,2\pi)$ parametrizing S_1. Thus $L^2(E_2 \otimes S_1)$ is the direct sum of the subspaces spanned by

(B.3) $$e^{-im\alpha}\, J_n(kr)\, e^{in\phi}\ .$$

But (5.1) is

(B.4) $$\alpha = -\phi - \psi$$

so that the basis

(B.5) $$\{e^{-im\alpha}\, J_n(kr)\, e^{in\phi} : k \in R^+,\ m,n \in Z\}$$

can be written as

(B.6) $$\{e^{im\psi}\, J_n(kr)\, e^{i(m+n)\phi} : k \in R^+,\ m,n \in Z\}$$

$$= \{e^{im\psi} J_{n-m}(kr) e^{in\phi} : k \in R^+,\ m,n \in Z\}$$

which is what we wished to prove.

REFERENCES

[1] R. BURRIDGE, Spherically symmetric differential equations, the rotation group, and tensor spherical functions, Proc. Camb. Phil. Soc. 65(1969), pp. 157-175.

[2] I.M. GELFAND, and Z. Ya. SHAPIRO, Representations of the group of rotations in three-dimensional space and their applications, Amer. Math. Soc. Transl. 2(1956), pp. 207-316.

[3] N. Ja. VILENKIN, Special Functions and the Theory of Group Representations, Translations of Mathematical Monographs, Vol. 22, American Mathematical Society, Providence, 1968.

GALERKIN METHODS FOR MISCIBLE DISPLACEMENT PROBLEMS WITH POINT SOURCES AND SINKS–UNIT MOBILITY RATIO CASE†

RICHARD E. EWING* AND MARY F. WHEELER**

Abstract. In most cases of reservoir simulation, the size of the well-bore is extremely small in comparison with the size of the reservoir. For this reason point sources and sinks are often used in mathematical models to describe the behavior at the wells. It can be shown that point singularities reduce the regularity of the functions which are determined by the mathematical models. In this paper we shall consider the problem of the miscible displacement of one incompressible fluid by another in a porous medium using point sources and sinks at the wells. Due to the reduced smoothness of the pressure and concentration caused by the assumption of point singularities, we must use a nonstandard analysis for the semidiscrete Galerkin procedures and obtain convergence rates of a lower order than for nonsingular sources and sinks. We treat only the case of unit mobility ratio, when the viscosities of the invading and displaced fluids are equal.

1. Introduction. In [7] the authors presented and analyzed certain numerical approximations by Galerkin methods for a problem arising in the miscible displacement of one incompressible fluid by another in a porous medium. Extensions of these methods to more efficient time-stepping procedures and more general boundary conditions [8], to interior penalty procedures [17], to methods of characteristics [13], to self-adaptive simulation techniques [6], and to mixed methods for pressure [5] have since been developed. These analyses were surveyed in [3]. All of the above analyses have made a major, and probably unphysical, assumption that the sources and sinks were smoothly distributed and the resulting functions of interest were thus fairly smooth in space. In this paper we shall present the first convergence analysis on this problem to appear in the literature where the sources and sinks are considered as point singularities or Dirac measures. The resulting pressure function thus has a finite number of logarithmic singularities located at the various wells. Since the

* Departments of Mathematics and Petroleum Engineering, University of Wyoming, Laramie, Wyoming 82071

** Department of Mathematical Sciences, Rice University, Houston, Texas 77001

† Research supported by Mobil Research and Development Corporation

resulting functions are considerably less smooth than previously assumed, the convergence rates obtained in this paper are slower than those previously obtained.

At present we are only able to analyze the special case where the viscosity of the invading fluid is equal to the viscosity of the resident fluid. In this case, the mobility ratio (see [5, 7]) is equal to one and the equation for pressure is a linear equation and can be uncoupled from the concentration equation and solved once for all time. Analysis for the general case when there is a nonlinear coupling between the pressure and concentration equations is in process and will appear elsewhere.

A set of model equations for our physical problem is given next. For a more detailed description of the physical problem, see [7, 11, 15]. Find the concentration $c = c(x, t)$ and $p = p(x, t)$ satisfying the following set of equations:

$$\text{(1.1)} \quad \text{a)} \quad \nabla \cdot [a(x)\{\nabla p - \gamma \nabla g\}] \equiv - \nabla \cdot u = \sum_{j=1}^{N} Q_j(t)\, \delta(x - x_j) \qquad , x \varepsilon \Omega,\ t \varepsilon J,$$

$$\text{b)} \quad \phi \frac{\partial c}{\partial t} + u \cdot \nabla c - \sum_{i=1}^{2} \sum_{j=1}^{2} \frac{\partial}{\partial x_i} \left[D_{ij}(x, u) \frac{\partial}{\partial x_j} c\right]$$

$$= \sum_{j=1}^{N} Q_j(t)(\tilde{c} - c)\, \delta(x - x_j) \qquad , x \varepsilon \Omega,\ t \varepsilon J,$$

with an initial condition and no flow boundary conditions given by

$$\text{(1.2)} \quad \text{a)} \quad c(x, 0) = c_0(x) \qquad , x \varepsilon \Omega,$$

$$\text{b)} \quad u \cdot \nu = 0 \qquad , x \varepsilon \partial\Omega,\ t \varepsilon J,$$

$$\text{c)} \quad \frac{\partial c}{\partial \nu} = 0 \qquad , x \varepsilon \partial\Omega,\ t \varepsilon J,$$

where Ω is a bounded domain in $\mathbb{R}^2$, $J = [0, T]$, and ν is the outward unit normal vector on $\partial\Omega$, the boundary of Ω. Here $a = a(x)$, $\gamma = \gamma(x)$, $g = g(x)$, $\phi = \phi(x)$ are specified reservoir and fluid properties, u is the Darcy velocity of the fluid, $\tilde{c}$ is the specified concentration at injection wells and the resident concentration at production wells, $\delta(x - x_j)$ is a Dirac delta function at $x = x_j$, and $Q_j(t)$ are the specified flow rates of the wells with the convention

(1.3) a) $Q_j(t) \geq 0$ for $j = 1, \ldots, N/2$ (injection wells),

b) $Q_j(t) \leq 0$ for $j = N/2 + 1, \ldots, N$ (production wells).

The diffusion tensor D_{ij} is given by

$$(1.4) \quad D(x, u) = \left(D_{ij}(x, u)\right) = \phi(x)\, D_0(x)\, I + \frac{\alpha_\ell}{|u|}\begin{pmatrix} u_1^2 & u_1u_2 \\ u_1u_2 & u_2^2 \end{pmatrix} + \frac{\alpha_t}{|u|}\begin{pmatrix} u_2^2 & -u_1u_2 \\ -u_1u_2 & u_1^2 \end{pmatrix}$$

where α_ℓ and α_t, the magnitudes of longitudinal and transverse dispersion, are given constants. Here for $v \in \mathbb{R}^2$, $|v|$ is the standard Euclidean norm of the vector. We make the physically realistic assumption on D_0, α_ℓ, and α_t that

$$(1.5) \quad 0 < D_* \, |\xi|^2 \leq \xi^T D(x,q)\, \xi, \qquad q \in \mathbb{R}^2,\ \xi \in \mathbb{R}^2 .$$

This gives us a coercivity property for the parabolic equation and an assumption of non-trivial diffusion and dispersion in the problem. We shall consider two separate cases for the diffusion tensor. In Case I $\alpha_\ell = \alpha_t = 0$ and we have only molecular diffusion while in Case II $\alpha_\ell > 0$ and $\alpha_t > 0$ help to model physical dispersion or mixing due to the flowing motion. Results for Cases I and II are presented in Theorems 3.1 and 3.2, respectively.

Using (1.1.a), we can see that pressure can be separated into its logarithmic singularity components and a smoother component, $\tilde{p}$, as follows:

$$(1.6) \quad p(x, t) = \sum_{i=1}^{N} \frac{Q_i(t)}{2\pi} \frac{1}{a(x_i)} \ln |x - x_i| + \tilde{p}(x,t) .$$

Similarly we can decompose the Darcy velocity as follows:

$$(1.7) \quad u(x, t) = \sum_{i=1}^{N} \frac{Q_i(t)}{2\pi} \frac{a(x)}{a(x_i)} \nabla \ln |x - x_i| + a(x)\, \nabla \tilde{p} .$$

We shall be more explicit about the smoothness of $\tilde{p}$ in Section 2 after we have presented the necessary terminology (see (2.8)). The fact that pressure is assumed to have logarithmic singularities also effects the smoothness of the concentration of the invading fluid. In particular, according to [14], $\int_0^T \int_\Omega \frac{\partial c^2}{\partial t}\, dxdt$ is not even bounded under the point sources assumption. Therefore the convergence analysis presented in this paper is non-standard and much more difficult than for the case of smoothly distributed sources and sinks.

The paper contains two additional sections. In Section 2, terminology is developed, basic regularity and boundedness assumptions are presented, basic projections needed for the analysis are considered, and the continuous-time Galerkin approximations of (1.1) and (1.2) are defined. In Section 3, a priori error estimates for the continuous-time approximations are obtained. L^2 rates of convergence for Cases I and II are given by $h^{1-\varepsilon}$ and $h^{1/2-\varepsilon}$, respectively.

2. Preliminaries and Description of the Galerkin Approximations. Let $(u, v) = \int_\Omega uv dx$ and $\|u\|^2 = (u, u)$ be the standard L^2 inner-product and norm. Let $W_s^k(\Omega)$ be the Sobolev space on Ω with norm

(2.1) $$\|\psi\|_{W_s^k} = \Big[\sum_{|\alpha| \leq k} \Big\| \frac{\partial^\alpha \psi}{\partial x^\alpha} \Big\|^s_{L^s(\Omega)} \Big]^{1/s},$$

with the usual modification for $s = \infty$. If $\nabla F = (F_1, F_2)$, write $\|\nabla F\|_{W_s^k}$ in place of $\big(\|F_1\|^s_{W_s^k} + \|F_2\|^s_{W_s^k}\big)^{1/s}$. When $s = 2$, denote $\|\psi\|_{W_2^k} \equiv \|\psi\|_{H^k} \equiv \|\psi\|_k$. If $k = 0$, $\|\psi\|_0 \equiv \|\psi\|$.

Let $\{M_h\}$ be a family of finite-dimensional subspaces of $H^1(\Omega)$ with the following property:

For $p = 2$ or ∞, there exist an integer $r \geq 2$ and a constant K_0 such that, for $1 \leq q \leq r$ and $\psi \varepsilon W_p^q(\Omega)$,

(2.2) a) $$\inf_{\chi \varepsilon M_h} \big\{\|\psi - \chi\|_{W_p^0} + h\|\psi - \chi\|_{W_p^1}\big\} \leq K_0 \|\psi\|_{W_p^q} h^q,$$

b) $$\inf_{\chi \varepsilon M_h} \big\{\|\psi - \chi\|_{W_\infty^0} + h\|\psi - \chi\|_{W_p^1}\big\} \leq K_0 \|\psi\|_q h^{q-1}.$$

We also define a family of finite-dimensional subspaces of $H^1(\Omega)$ called $\{N_h\}$ which satisfies the same property as $\{M_h\}$ with r replaced by s. We also assume that the families $\{M_h\}$ and $\{N_h\}$ satisfy the following so-called "inverse hypotheses":

if $\psi \varepsilon M_h$ or N_h, for some $K_1 > 0$,

(2.3) a) $\|\psi\|_{L^p} \leq K_1 h^{\frac{2}{p}-1} \|\psi\|$, $2 \leq p \leq \infty$

b) $\|\psi\|_1 \leq K_1 h^{-1} \|\psi\|$.

We shall use M_h to approximate c and N_h to approximate the non-logarithmic part of p.

We shall make the same boundedness assumptions and somewhat weaker smoothness assumptions on the coefficients than were made in [7, 8, 17]. We consider spaces of the form

$$\|\psi\|_{W_p^q((a,b),X)} = \left\{\psi : (a, b) \longrightarrow X \mid \| \frac{\partial^\alpha \psi}{\partial t^\alpha} (t)\|_X \, \varepsilon \, L^p((a, b))\right\}$$

with norm

$$(2.4) \qquad \|\psi\|_{W_p^q((a,b),X)} = \left[\sum_{|\alpha| \leq q} \| \, \| \frac{\partial^\alpha \psi}{\partial t^\alpha} (t)\|_X \, \|^p_{L^p(a,b)} \right]^{1/p} ,$$

where $1 \leq p, q \leq \infty$ and X is a Sobolev space in our applications. When (a, b) = J, we shall suppress (a, b) in our notation in (2.4). Let (p, c), the solution of (1.1)-(1.2), satisfy the following regularity assumptions:

(2.5) a) $\|c\|_{L^2(H^{2-\varepsilon})} + \|c\|_{L^2(W_p^1)} + \|c\|_{L^\infty(H^{1-\varepsilon})} \leq K_2$,

b) $\|p\|_{L^\infty(H^{1-\varepsilon})} \leq K_2$,

c) $\|u\|_{L^\infty(L^{2-\varepsilon})} \leq K_2$,

d) $\|\frac{\partial c}{\partial t}\|_{L^2(L^{2-\varepsilon})} \leq K_2$,

where $\varepsilon > 0$ can be chosen arbitrarily small, p can be chosen arbitrarily large, $K_2 > 0$ is a fixed constant, and J has been supressed in the index notation of the norms. These regularity assumptions are based on analysis by Sammon [14].

In our analysis we shall use two different approximations for c from M_h. We first define the L^2 projection $\hat{c}$ of c into M_h by

$$(2.6)\qquad a)\ \ (\phi(c - \hat{c}), \chi) = 0\ , \qquad \chi \varepsilon\ M_h\ ,\ \text{or}$$

$$b)\ \ (\phi(\frac{\partial c}{\partial t} - \frac{\partial \hat{c}}{\partial t}), \chi = 0\ , \qquad \chi \varepsilon\ M_h\ .$$

We are led to use the L^2 projection of c into M_h instead of the now more standard H^1 projection due to smoothness restrictions on c. Since we assume that $\frac{\partial c}{\partial t}$ is only in $L^2(J, L^{2-\varepsilon})$ for ε arbitrarily small, we are not able to treat terms like $\frac{\partial}{\partial t}(c - \hat{c})$ in a normal fashion. Thus we have used $\hat{c}$ to project this problem away as in (2.6.b). This causes reduced accuracy in terms like $\nabla(c - \hat{c})$, but the loss of accuracy was inevitable in any case due to the logarithmic singularities in pressure. We also denote by c_I the interpolant of c in M_h. We then use (2.3) and the theory of interpolation spaces to obtain the following approximation theory results:

Lemma 2.1: There exists a positive constant $K_2 = K_2(\Omega, K_0, K_2)$ such that, for each $t \varepsilon J$ and ε arbitrarily small,

$$(2.7)\qquad a)\ \ \|c - \hat{c}\| + h\|c - \hat{c}\|_1 \leq K_2\|c\|_{q_1}\ h^{q_1} \qquad , 0 \leq q_1 \leq 2 - \varepsilon,$$

$$b)\ \ \|c - \hat{c}\|_{L^\infty} \leq K_2\|c\|_{W_\infty^{q_2}}\ h^{q_2} \qquad , 0 \leq q_2 \leq 1 - \varepsilon,$$

$$c)\ \ \|c - c_I\| + h[\|c - c_I\|_1 + \|c - c_I\|_{L^\infty}]$$

$$\leq K_2\|c\|_{q_3}\ h^{q_3} \qquad , 1 \leq q_3 \leq 2 - \varepsilon.$$

Proof: See [2, 4, 16].

We next note that from [9] we know that $\tilde{p}$ defined in (1.6) satisfies

$$(2.8)\qquad \|\tilde{p}\|_{H^{2-\varepsilon}} \leq K_3$$

for $K_3 > 0$ and some arbitrarily small $\varepsilon > 0$. We shall use the logarithmic part of p defined in (1.6) to form the leading part of our approximation P of p. We define P to be

$$(2.9)\qquad P(x,t) = \sum_{i=1}^{N} \frac{Q_i(t)}{a(x_i)}\,\ell n\ |x - x_i| + \tilde{P}\ ,$$

where x_i, $i=1, \ldots, N$, are the locations of the injection and production wells and $\tilde{P} \in N_h$ is an approximation to $\tilde{p}$ from (1.6) defined for each $t \in J$ by

$$(2.10) \qquad \left(a(\cdot)\, \nabla(\tilde{p} - \tilde{P}),\, \nabla \chi\right) = 0 , \qquad \chi \in M_h .$$

This is an example of a weighted elliptic projection used by Wheeler in [16]. We obtain the following result.

Lemma 2.2: There exists a positive constant $K_4 = K_4(\Omega, K_0, K_2)$ such that, for each $t \in J$ and ε arbitrarily small,

$$(2.11) \qquad \|\nabla(\tilde{p} - \tilde{P})\| \leq K_4 \|p\|_{H^q} h^{q-1} ,$$

for $1 \leq q \leq 2 - \varepsilon$.

If we then define u and U by,

$$(2.12) \qquad \text{a)} \quad u = a(x)\, \nabla p = \sum_{j=1}^{N} Q_j(t) \frac{a(x)}{a(x_j)} \nabla \ln|x - x_j| + a(x)\, \nabla \tilde{p} ,$$

$$\text{b)} \quad U = a(x)\, \nabla P = \sum_{j=1}^{N} Q_j(t) \frac{a(x)}{a(x_j)} \nabla \ln|x - x_j| + a(x)\, \nabla \tilde{P} ,$$

we can immediately use (2.8) and (2.11) to obtain for each $t \in J$ and $K_5 = K_5(K_3, K_4, a(x))$

$$(2.13) \qquad \|u - U\| \leq K_5\, h^{1-\varepsilon}$$

where $\varepsilon > 0$ can be made arbitrarily small.

We next define the continuous-time approximation of c as follows: let $C:[0,T] \longrightarrow M_h$ be defined by

$$(2.14) \qquad \text{a)} \quad \left(\phi \frac{\partial C}{\partial t}, \chi\right) + \sum_{i=1}^{2} \sum_{j=1}^{2} \left(D_{ij}(U) \frac{\partial}{\partial x_j} C, \frac{\partial}{\partial x_i} \chi\right) + (U \cdot \nabla C, \chi)$$

$$= \sum_{j=1}^{N/2} Q_j(t)\, (\tilde{c} - C)(x_j, t)\, \chi(x_j) \qquad , \chi \in M_h ,$$

$$\text{b)} \quad \left(C(0) - c_0, \chi\right) = 0 \qquad , \chi \in M_h ,$$

where U, P, and $\tilde{P}$ are defined by (2.12.b), (2.9), and (2.10), respectively. The main results of this paper are a priori estimates for the error in approximating c from (1.1) by C from (2.14). These will appear in the next section.

3. A Priori Error Estimates. In this section, we shall obtain a priori bounds for the error in the concentration approximation C - c, to go with the bound of the error in the Darcy velocity approximation given by (2.13). We shall split our a priori estimates into two cases. Case I will reflect the assumption that the only diffusion present in the model is molecular diffusion and $\alpha_\ell = \alpha_t = 0$ in (1.4). Case II will extend the estimates to the more difficult case of tensorial physical dispersion given by (1.4) with $\alpha_\ell > 0$ and $\alpha_t > 0$. As expected, we obtain a reduced convergence rate for the more difficult case.

Theorem 3.1. Let (c, p) satisfy (1.1)-(1.2) and (C, P) satisfy (2.9), (2.10), and (2.14). For the molecular diffusion case, let $\alpha_\ell = \alpha_t = 0$ in (1.4). There exist positive constants $K_6 = K_6\big(\Omega, K_i, i=0 \ldots, 5\big)$ and h_0 such that, if $h \leq h_0$,

$$(3.1) \qquad \|c - C\|_{L^\infty(J,L^2)} + \|\nabla(c - C)\|_{L^2(J,L^2)}$$

$$+ \left\{ \sum_{j=1}^{N} \int_0^T |Q_j(t)|(C - c)^2(x_j, t)\, dt \right\}^{\frac{1}{2}} \leq K_6\, h^{1-\hat{\varepsilon}} .$$

($\hat{\varepsilon} = \hat{\varepsilon}(\varepsilon) > 0$ is defined in (3.19) below and can be taken arbitrarily small).

Proof: Let $\xi = C - \hat{c}$ and $\eta = c - \hat{c}$ with $\hat{c}$ from (2.6) and C from (2.14). Subtract (2.6) from (2.14) and let $\chi = \xi$ to obtain

$$(3.2) \qquad \Big(\phi \frac{\partial \xi}{\partial t}, \xi\Big) + (\phi\, D_0\, \nabla\xi, \nabla\xi) + (u \cdot \nabla\, \xi, \xi)$$

$$= (\phi\, D_0\, \nabla\, \eta, \nabla\, \xi) + (u \cdot \nabla\, \eta, \xi) + ((u - U) \cdot \nabla\, C, \xi)$$

$$+ \sum_{j=1}^{N/2} Q_j(t)(c - C)(x_j, t)\, \xi(x_j, t) ,$$

For the last term on the left-hand side of (3.2), we integrate by parts (note that $\frac{\partial u}{\partial \nu} = 0$ on $\partial\Omega$) and use (1.1.a) with $\chi = \xi$ to obtain

$$(3.3) \qquad (u \cdot \nabla \xi, \xi) = \left(u \cdot \nabla \frac{\xi^2}{2}, 1\right)$$

$$= -\left(\nabla \cdot u, \frac{\xi^2}{2}\right)$$

$$= -\frac{1}{2} \sum_{j=1}^{N} Q_j(t)\, \xi^2(x_j, t)$$

$$= -\frac{1}{2} \sum_{j=1}^{N/2} |Q_j(t)|\, \xi^2(x_j, t) + \frac{1}{2} \sum_{j=N/2+1}^{N} |Q_j(t)|\, \xi^2(x_j, t).$$

We then combine part of the last term on the right side of (3.2) with (3.3) and replace c by c_I at the wells, to obtain

$$(3.4) \qquad \frac{1}{2} \frac{d}{dt} \|\phi^{1/2} \xi\|^2 + \|(\phi\, D_0)^{1/2} \nabla \xi\|^2 + \frac{1}{2} \sum_{j=1}^{N} |Q_j(t)|\, \xi^2(x_j, t)$$

$$= (\phi\, D_0 \nabla \eta, \nabla \xi) + (u \cdot \nabla \eta, \xi) + ((u - U) \cdot \nabla C, \xi)$$

$$+ \sum_{j=1}^{N/2} Q_j(t)(c_I - \hat{c})(x_j, t)\, \xi(x_j, t)$$

$$= T_1 + T_2 + T_3 + T_4 .$$

We next integrate (3.4) termwise on τ in $J_t = [0, t]$ for $t \varepsilon J$. The left-hand side of the resulting equation is then bounded below as follows

$$(3.5) \qquad \frac{1}{2} \int_0^t \frac{d}{d\tau} \|\phi^{1/2} \xi\|^2\, d\tau + \int_0^t \|(\phi\, D_0)^{1/2} \nabla \xi\|^2\, d\tau$$

$$+ \frac{1}{2} \sum_{j=1}^{N} \int_0^t |Q_j(\tau)|\, \xi^2(x_j, \tau)\, d\tau$$

$$\geqslant \alpha\Big[\|\xi(t)\|^2 + \|\nabla \xi\|^2_{L^2(J_t, L^2)} + \sum_{j=1}^{N} \int_0^t |Q_j(\tau)|\, \xi^2(x_j, \tau)\, d\tau\Big]$$

where α depends upon uniform lower bounds for the coefficients ϕ and D_0 such as D_* from (1.5). We next consider bounds for the terms on the right-hand side of the integrated analogue of (3.4). We note that from (2.7.a), we obtain

$$(3.6)\qquad \left|\int_0^t T_1\,d\tau\right| \leq K\int_0^t \|\nabla\eta\|\,\|\nabla\xi\|\,d\tau \leq K\,h^{1-\varepsilon}\,\|\nabla\xi\|_{L^2(J_t,L^2)}$$

$$\leq \frac{\alpha}{8}\,\|\nabla\xi\|_{L^2(J_t,L^2)} + K\,h^{2(1-\varepsilon)}\ ,$$

where α is from (3.5) and K is used here and in the following as a generic positive constant, usually of different size with each use. Then using (2.3.a), (2.5.c), (2.7.a) and the fact [1, 10] that, for $\Omega \subset \mathbb{R}^2$ and for any $1 \leq p < \infty$,

$$(3.7)\qquad \|\chi\|_{L^p} \leq K\,\|\chi\|_1\ ,$$

we use the Sobolev Imbedding Theorem [1, 10] to see that

$$(3.8)\qquad \left|\int_0^t T_2\,d\tau\right| \leq \|u\|_{L^\infty(L^{2-\varepsilon})}\int_0^t \|\nabla\eta\|_{L^{2+\varepsilon_1}}\,\|\xi\|_{L^p}\,d\tau$$

$$\leq K_2\,h^{-\frac{\varepsilon_1}{2+\varepsilon_1}}\,K_1\,\|\nabla\eta\|_{L^2(L^2)}\,\|\xi\|_{L^2(J_t,H^1)}$$

$$\leq K\,h^{1-\varepsilon-\frac{\varepsilon_1}{2+\varepsilon_1}}\left[\|\xi\|_{L^2(J_t,L^2)} + \|\nabla\xi\|_{L^2(J_t,L^2)}\right]$$

$$\leq \frac{\alpha}{8}\left(\|\nabla\xi\|^2_{L^2(J_t,L^2)}\ \|\xi\|^2_{L^2(J_t,L^2)}\right) + K\,h^{2\left(1-\varepsilon-\frac{\varepsilon_1}{2+\varepsilon_1}\right)}$$

where ε, ε_1, and p satisfy the relation

$$(3.9)\qquad \frac{1}{2-\varepsilon} + \frac{1}{2+\varepsilon_1} + \frac{1}{p} = 1\ .$$

We note that since $\varepsilon > 0$ from (2.5.c) can be arbitrarily small and arbitrarily large p satisfies (3.7), ε_1 can also be taken arbitrarily small and still satisfy (3.9). We next use (2.7.b–c) to obtain

$$(3.10)\qquad \left|\int_0^t T_4\,d\tau\right| \leq \frac{\alpha}{8}\sum_{j=1}^{N}\int_0^t |Q_j(\tau)|\,\xi^2(x_j,\tau)\,d\tau + K\,h^{2(1-\varepsilon)}\ .$$

Finally we shall break T_3 into pieces to consider as follows:

$$(3.11)\qquad T_3 = ((u-U)\cdot\nabla\xi,\xi) - ((u-U)\cdot\nabla\eta,\xi) + ((u-U)\cdot\nabla c,\xi) = T_5 + T_6 + T_7 \ .$$

Now we again use the Sobolev Imbedding Theorem with $\varepsilon_2 > 0$ arbitrarily small and $p_2 > 0$ arbitrarily large satisfying

$$(3.12)\qquad \frac{1}{2} + \frac{1}{2+\varepsilon_2} + \frac{1}{p_2} = 1$$

and apply (2.3.a), (2.13), and (3.7) to obtain

$$\begin{aligned}(3.13)\qquad \left|\int_0^t T_5\,d\tau\right| &\le \|u-U\|_{L^\infty(L^2)}\int_0^t \|\nabla\xi\|_{L^{2+\varepsilon_2}}\|\xi\|_{L^{p_2}}\,d\tau \\ &\le K_5 h^{1-\varepsilon} K_1 h^{-\frac{\varepsilon_2}{2+\varepsilon_2}} \|\nabla\xi\|_{L^2(J_t,L^2)}\|\xi\|_{L^2(J_t,H^1)} \\ &\le K h^{1/2}\|\nabla\xi\|_{L^2(J_t,L^2)}\left[\|\xi\|_{L^2(J_t,L^2)} + \|\nabla\xi\|_{L^2(J_t,L^2)}\right] \\ &\le \frac{\alpha}{8}\left[\|\nabla\xi\|^2_{L^2(J_t,L^2)} + \|\xi\|^2_{L^2(J_t,L^2)}\right] .\end{aligned}$$

In (3.13) we have chosen ε and ε_2 sufficiently small that $\varepsilon + \frac{\varepsilon_2}{2+\varepsilon_2} < \frac{1}{2}$. In the same fashion, we use (2.3), (2.7a), and (2.13) to see that

$$\begin{aligned}(3.14)\qquad \left|\int_0^t T_6\,d\tau\right| &\le \|u-U\|_{L^\infty(L^2)}\int_0^t \|\nabla\eta\|\,\|\xi\|_{L^\infty}\,d\tau \\ &\le K_5 h^{1-\varepsilon} K_2 h^{1-\varepsilon} K_1 h^{-1}\|\xi\|_{L^2(J_t,L^2)} \\ &\le \frac{\alpha}{8}\|\xi\|^2_{L^2(J_t,L^2)} + K h^{2(1-2\varepsilon)} .\end{aligned}$$

Now since (2.5.a) holds for $\varepsilon > 0$ arbitrarily small, an imbedding result similar to the one used in (3.7) can be applied to pick a $p_3 = p_3(\varepsilon) > 0$, arbitrarily large, and satisfying

$$(3.15) \qquad \|\nabla c\|_{L^2(L^{p_3})} \leq K_2$$

from (2.5.a). Using this p_3, we choose $\varepsilon_3 = \varepsilon_3(p_3, \varepsilon) > 0$, arbitrarily small, to satisfy

$$(3.16) \qquad \frac{1}{2} + \frac{1}{p_3} + \frac{1}{2 + \varepsilon_3} = 1 .$$

Then we see that, as before,

$$(3.17) \qquad \left|\int_0^t T_7 \, d\tau\right| \leq \|u - U\|_{L^\infty(L^2)} \int_0^t \|\nabla c\|_{L^{p_3}} \|\xi\|_{L^{2+\varepsilon_3}} \, d\tau$$

$$\leq K_5 \, h^{1-\varepsilon} \|\nabla c\|_{L^2(L^{p_3})} K_1 \, h^{-\frac{\varepsilon_3}{2+\varepsilon_3}} \|\xi\|_{L^2(J_t, L^2)}$$

$$\leq \frac{\alpha}{8} \|\xi\|^2_{L^2(J_t,L^2)} + K \, h^{2\left(1-\varepsilon-\frac{\varepsilon_3}{2+\varepsilon_3}\right)} .$$

We next combine the above estimates to see that for each $t \varepsilon [0, T]$, we have

$$(3.18) \qquad \|\xi(t)\|^2 + \|\nabla \xi\|^2_{L^2(J_t,L^2)} + \sum_{j=1}^{N} \int_0^t |Q_j(\tau)| \, \xi^2(x_j, \tau) \, d\tau$$

$$\leq \|\xi\|^2_{L^2(J_t,L^2)} + K \, h^{2(1-\hat{\varepsilon})}$$

where $\hat{\varepsilon}$ is defined as

$$(3.19)\qquad \hat{\varepsilon} = \max\left[2\varepsilon,\ \varepsilon + \frac{\varepsilon_1}{2+\varepsilon_1},\ \varepsilon + \frac{\varepsilon_3}{2+\varepsilon_3}\right]$$

and can be taken arbitrarily small. We can now apply Gronwall's lemma to (3.18) and use (2.7) and the triangle inequality to obtain the desired result (3.1).

We next consider Case II where $\alpha_\ell > 0$ and $\alpha_t > 0$ model physical dispersion [12]. We obtain a reduced rate of convergence in this more complex case.

Theorem 3.2. Let (c, p) satisfy (1.1)-(1.2) and (C, P) satisfy (2.9), (2.10), and (2.14). There exist positive constants $K_7 = K_7(\Omega,\ K_i;\ i=0,\ \ldots,\ 5)$ and h_0 such that, if $h \le h_0$,

$$(3.20)\qquad \|c - C\|_{L^\infty(J,L^2)} + \|\nabla(c - C)\|_{L^2(J,L^2)}$$

$$+ \left\{\sum_{j=1}^{N} \int_0^T |Q_j(t)|(C - c)^2(x_j,\ t)\ dt\right\}^{1/2} \le K_7\ h^{1/2-\bar{\varepsilon}} .$$

($\bar{\varepsilon} = \bar{\varepsilon}(\varepsilon) > 0$ is defined in (3.35) below and can be taken arbitrarily small.)

Proof: Let ξ and η be as in the proof of Theorem 3.1. Subtracting (2.6) from (2.14) in this Case II and substituting ξ for χ yields

$$(3.21)\qquad \left(\phi \frac{\partial \xi}{\partial t},\ \xi\right) + \sum_{i=1}^{2} \sum_{j=1}^{2} \left(D_{ij}(U) \frac{\partial}{\partial x_j} \xi,\ \frac{\partial}{\partial x_i} \xi\right) + (u \cdot \nabla \xi,\ \xi)$$

$$= (u \cdot \nabla \eta,\ \xi) + ((u - U) \cdot \nabla C,\ \xi)$$

$$+ \sum_{j=1}^{N/2} Q_j(t)(c - C)(x_j,\ t)\ \xi(x_j,\ t)$$

$$+ \sum_{i=1}^{2} \sum_{j=1}^{2} \left(D_{ij}(u) \frac{\partial}{\partial x_j} c - D_{ij}(U) \frac{\partial}{\partial x_j} \hat{c},\ \frac{\partial}{\partial x_i} \xi\right)$$

$$\equiv T_8 + T_9 + T_{10} + T_{11} .$$

We again integrate (3.21) termwise on τ in $J_t = [0, t]$ for $t\varepsilon J$. We obtain an analogue of (3.5) for the left-hand side of (3.21) where now α depends upon the constant D_* assumed in (1.5). All the terms are then treated exactly as in the proof of Theorem 3.1 except for T_{11} which did not appear in Case I. We first split T_{11} up as follows:

$$(3.22)\qquad T_{11} = \sum_{i=1}^{2}\sum_{j=1}^{2}\left(\left[D_{ij}(u) - D_{ij}(U)\right]\frac{\partial}{\partial x_j}c,\ \frac{\partial}{\partial x_i}\xi\right) - \sum_{i=1}^{2}\sum_{j=1}^{2}\left(D_{ij}(U)\frac{\partial}{\partial x_j}\eta,\ \frac{\partial}{\partial x_i}\xi\right) \equiv T_{12} + T_{13}.$$

We first note that

$$(3.23)\qquad \left|\int_0^t T_{13}\,d\tau\right| \leq 4\sum_{i=1}^{2}\sum_{j=1}^{2}\int_0^t\left(D_{ij}(U)\frac{\partial}{\partial x_j}\eta,\ \frac{\partial}{\partial x_i}\eta\right)d\tau + \frac{1}{4}\sum_{i=1}^{2}\sum_{j=1}^{2}\int_0^t\left(D_{ij}(U)\frac{\partial}{\partial x_j}\xi,\ \frac{\partial}{\partial x_i}\xi\right)d\tau \equiv T_{14} + T_{15}.$$

Clearly, T_{15} can be subtracted from the corresponding term on the left-hand side of (3.21). We can then split T_{14} as follows:

$$(3.24)\qquad |T_{14}| \leq 4\sum_{i=1}^{2}\sum_{j=1}^{2}\int_0^t\left(\left[D_{ij}(U) - D_{ij}(u)\right]\frac{\partial}{\partial x_j}\eta,\ \frac{\partial}{\partial x_i}\eta\right)d\tau + 4\sum_{i=1}^{2}\sum_{j=1}^{2}\int_0^t\left(D_{ij}(u)\frac{\partial}{\partial x_j}\eta,\ \frac{\partial}{\partial x_i}\eta\right)d\tau \equiv T_{16} + T_{17}.$$

In order to bound $|\int_0^t T_{12}\, d\tau + T_{16}|$, we shall use (3.15) and an analogue for η. First $c - c_I$ satisfies a bound of the form

$$(3.25)\qquad \|\nabla(c - c_I)\|_{L^2(L^{p_3})} \le K$$

where p_3 is the same as in (3.15). Then since $c_I - \hat{c} \in M_h$, we can use (2.3), (2.5), and (2.7) to see that, for $\varepsilon < \frac{2}{p_3}$,

$$(3.26)\qquad \|\nabla(c_I - \hat{c})\|_{L^2(L^{p_3})} \le K_1\, h^{\frac{2}{p_3}-1} \|\nabla(c_I - \overset{\Delta}{c})\|_{L^2(L^2)}$$

$$\le K_1\, h^{\frac{2}{p_3}-1} K_2\, h^{1-\varepsilon} \|c\|_{L^2(H^{2-\varepsilon})}$$

$$\le K$$

Combining (3.25) and (3.26), we have

$$(3.27)\qquad \|\nabla \eta\|_{L^2(L^{p_3})} \le K .$$

We next note that by elementary but tedious computations one can show that $D_{ij}(x,u)$ is Lipschitz in u with Lipschitz constant 3. Thus combining (3.15) and (3.27) and the Lipschitz behavior of D_{ij}, we can use (2.3.a) and (2.13) to obtain

$$(3.28)\qquad |\int_0^t T_{12}\, d\tau + T_{16}|$$

$$\le \|U - u\|_{L^\infty(L^2)} \int_0^t \left[\|\nabla \eta\|_{L^{p_3}} + \|\nabla c\|_{L^{p_3}}\right] \|\nabla \xi\|_{L^{2+\varepsilon_3}}\, d\tau$$

$$\le K_5\, h^{1-\varepsilon}\, K\, K_1\, h^{-\frac{\varepsilon_3}{2+\varepsilon_3}} \|\nabla \xi\|_{L^2(J_t, L^2)}$$

$$\le \frac{\alpha}{8} \|\nabla \xi\|^2_{L^2(J_t,L^2)} + K\, h^{2\left(1-\varepsilon-\frac{\varepsilon_3}{2+\varepsilon_3}\right)} .$$

We next use (1.7) and (2.12.a) to note that for each i, j = 1, 2,

$$(3.29) \qquad |D_{ij}(u)| \leq |u| \leq \sum_{k=1}^{N} K_8 \frac{1}{|x - x_k|} + |a(x) \nabla \tilde{p}|$$

where $|v|$ for a vector v is the standard Euclidean norm in $\mathbb{R}^2$. Using (2.3), (2.7), (2.8) and (3.29), we see that

$$(3.30) \qquad |T_{17}| \leq |T_{18}| + \|\nabla \tilde{p}\|_{L^\infty(L^4)} \int_0^t \|\nabla \eta\|_{L^4} \|\nabla \eta\| \, d\tau$$

$$\leq |T_{18}| + K h^{-\frac{1}{2}} \|\nabla \eta\|^2_{L^2(L^2)}$$

$$\leq |T_{18}| + K h^{-\frac{1}{2}} h^{2(1-\varepsilon)}$$

$$\leq |T_{18}| + K h^{2(\frac{3}{4} - \varepsilon)} .$$

We note that the second term on the right side of (3.30) does not give an optimal estimate for the $a(x)\nabla\tilde{p}$ term from (3.29), but yields a better bound than we are at present able to obtain for T_{18}. T_{18} contains a term of size $|x - x_j|^{-1}$ centered at each well. For simplicity, we will carefully estimate only one such term. Without loss of generality assume we have a well centered at the origin x = 0. We shall then split this term by considering the spatial integration over B_h^β, a disc of radius h^β centered at the origin, and its complement $\Omega - B_h^\beta$. We then see that

$$(3.31) \qquad |T_{18}| \leq N K \sum_{i=1}^{2} \sum_{j=1}^{2} \int_0^t \int_{B_h^\beta} \frac{1}{r} \frac{\partial \eta}{\partial x_j} \frac{\partial \eta}{\partial x_i} \, dx d\tau$$

$$+ N K \sum_{i=1}^{2} \sum_{j=1}^{2} \int_0^t \int_{\Omega - B_h^\beta} \frac{1}{r} \frac{\partial \eta}{\partial x_j} \frac{\partial \eta}{\partial x_i} \, dx d\tau$$

$$\equiv T_{19} + T_{20} .$$

we then obtain

$$(3.32)\qquad |T_{20}| \leq K\, h^{-\beta} \, \|\nabla \eta\|^2_{L^2(L^2)}$$

$$\leq K\, h^{-\beta}\, h^{2(1-\varepsilon)} \equiv K\, h^{2-2\varepsilon-\beta} .$$

Next, we let p_3 and q satisfy

$$(3.33)\qquad \frac{2}{p_3} + \frac{1}{q} = 1$$

and use (3.27) to obtain

$$(3.34)\qquad |T_{19}| \leq K\, \|\nabla \eta\|^2_{L^2(L^{p_3})} \int_0^{2\pi} \Big(\int_0^{h^\beta} r^{-q+1}\, dr\Big)^{\frac{1}{q}} d\theta$$

$$\leq K(h^\beta)^{\frac{-q+2}{q}} = K(h^\beta)^{1-\frac{4}{p_3}}$$

$$= K\, h^{\beta-\frac{4\beta}{p_3}}$$

for p_3 arbitrarily large. We then pick β to balance (3.32) and (3.34). With $\beta = 1$ and

$$(3.35)\qquad \bar{\varepsilon} = \max\Big[\varepsilon, \frac{2}{p_3}\Big],$$

we see that

$$(3.36)\qquad |T_{18}| \leq K\, h^{2(\frac{1}{2}+\bar{\varepsilon})} .$$

Combining the above estimates and corresponding bounds from the proof of Theorem 3.1, we obtain, for each $t\varepsilon[0, T]$.

$$(3.37) \qquad \|\xi(t)\|^2 + \|\nabla \xi\|^2_{L^2(J_t, L^2)} + \sum_{j=1}^{N} \int_0^t |Q_j(\tau)|\, \xi^2(x_j, \tau)\, d\tau$$

$$\leq \|\xi\|^2_{L^2(J_t, L^2)} + K\, h^{2(\frac{1}{2} - \bar{\varepsilon})}$$

where $\bar{\varepsilon}$ is given by (3.36) and can be taken arbitrarily small. Then applying Gronwall's lemma to (3.37) we use (2.7) and the triangle inequality to obtain the desired result, (3.20).

REFERENCES

[1] R. A. ADAMS, Sobolev Spaces, Academic Press, New York, 1975.

[2] J. H. BRAMBLE and S. R. HILBERT, Bounds for a class of linear functionals with applications to Hermite interpolation, Numer. Math. 16, pp. 362-369.

[3] J. DOUGLAS, JR., The numerical solution of miscible displacement in porous media, Computational Methods in Nonlinear Mechanics, (J. T. Oden, ed.), North Holland, New York, 1980.

[4] J. DOUGLAS, JR., T. DUPONT, and R. E. EWING, Incomplete iteration for time-stepping a Galerkin method for a quasilinear parabolic problem, SIAM J. Numer. Anal. 16 (1979), pp. 503-522.

[5] J. DOUGLAS, JR., R. E. EWING, and M. F. WHEELER, Mixed methods for miscible displacement problems in porous media, (to appear).

[6] J. DOUGLAS, JR., M. F. WHEELER, B. L. DARLOW, and R. P. KENDALL, Self-adaptive finite element simulation of miscible displacement in porous media, SIAM J. Sci. Stat. Comp. (to appear).

[7] R. E. EWING and M. F. WHEELER, Galerkin methods for miscible displacement problems in porous media, SIAM J. Numer. Anal. 17 (1980), pp. 351-365.

[8] R. E. EWING and T. F. RUSSELL, Efficient time-stepping procedures for miscible displacement problems in porous media, SIAM J. Numer. Anal. (to appear).

[9] P. GRISVARD (private communication).

[10] J. L. LIONS and E. MAGENES, Non-homogeneous Boundary Value Problems and Applications, Vol. I, Springer-Verlag, New York, 1972.

[11] D. W. PEACEMAN, Fundamentals of Numerical Reservoir Simulation, Elsevier Publishing Company, 1977.

[12] D. W. PEACEMAN, Improved treatment of dispersion in numerical calculation of multidimensional miscible displacement, Soc. Pet. Eng. J. (1966), pp. 213-216.

[13] T. F. RUSSELL, An incompletely iterated characteristic finite element method for a miscible displacement problem, Ph.D. Thesis, University of Chicago, Chicago, 1980.

[14] P. H. SAMMON (private communication).

[15] A. SETTARI, H. S. PRICE, and T. DUPONT, Development and application of variational methods for simulation of miscible displacement in porous media, Soc. Pet. Eng. J. (June 1977), pp. 228-246.

[16] M. F. WHEELER, A priori L^2-error estimates for Galerkin approximations to parabolic partial differential equations, SIAM J. Numer. Anal. 10 (1973), pp. 723-759.

[17] M. F. WHEELER and B. L. DARLOW, Interior penalty Galerkin methods for miscible displacement problems in porous media, Computational Methods in Non-linear Mechanics, J. T. Oden, editor, North Holland Publishing Company, New York (1980), pp. 485-506.

THE METHOD OF LINES, LINE SOR AND FREE BOUNDARIES

GUNTER H. MEYER*

Abstract. The method of lines for elliptic problems is reviewed. When the problem is imbedded into a parabolic problem and the method of lines with continuous time is applied then an initial value problem for ordinary differential equations results. A non-standard explicit numerical integrator then leads the usual elliptic SOR method. When the method of lines is applied directly to the elliptic equation a boundary value problem for a system of ordinary differential equations results. It is shown that the iterative solution of this system with the sweep method using an implicit/explicit Euler method is exactly the line SOR method for the fully discretized elliptic equation. The value of this approach for free boundary problems is illustrated by solving the classical porous dam problem.

1. Introduction. Boundary value problems for elliptic and parabolic equations are usually solved by reducing them to algebraic sytems through finite differences or finite elements. Various solution methods then exist to solve the algebraic problem efficiently. On occasions, the original problem is only partially discretized so that a system of differential equations rather than algebraic equations results. Such solution method is commonly called the method of lines and as such has received considerable attention in the research literature. In the end, however, the system of differential equations needs to be solved. If a further discretization is carried out then again a fully discrete method for the original problem is obtained and it becomes natural to ask whether the approach via the method of lines presents anything new and useful.

It is the purpose of this paper to examine this question in connection with elliptic boundary value problems. We shall first discuss the imbedding of the elliptic problem into a parabolic problem to which the time-continuous method of lines can be applied. We shall find that the classical point SOR method for the elliptic problem may be interpreted as an integrator for the method of lines.

*School of Mathematics, Georgia Institute of Technology, Atlanta, Georgia 30332
Research supported under NSF Grant MCS 8302547

Some estimates for the optimum relaxation factor can be obtained with Fourier stability arguments.

We shall then describe the more common method of lines for elliptic boundary value problems where the partial differential equation is approximated by a system of ordinary differential equations subject to boundary data. The system is solved iteratively by a sweep method which in a special case is shown to be the usual line SOR method for the five point (discrete) Laplace operator. The usefulness of this approach for irregular domains and free boundaries will be pointed out. The classical dam problem and a reaction-diffusion system will be used to illustrate the application of the method of lines for elliptic problems.

2. Parabolic Problems and Elliptic SOR. In the numerical solution of parabolic initial/boundary value problems it is common to separate the space and time discretizations. After space is discretized through finite elements, collocation, or finite differences an initial value problem for a system of ordinary differential equations results. This system is usually called the (continuous time) method of lines approximation for the parabolic problem. The method of lines for parabolic equations has been extensively analyzed and numerical integration techniques have been implemented in production computer codes.

It is well known that the solution to an elliptic boundary value problem may be thought of as the equilibrium solution of a related parabolic initial/boundary value problem [4, p. 156]. In principle one can integrate the method of lines equations with currently available codes in order to find the steady state solution. However, the initial value problem for ordinary differential equations which approximates the parabolic problem is stiff. As a rule this requires implicit integrators so that the effort expended per time step is comparable to a direct solution of the whole elliptic problem. Hence this method of lines appears of little practical use for elliptic problems.

On the other hand, the imbedding into a parabolic problem furnishes a new way of interpreting iterative methods for elliptic equations. We would like to illustrate this point.

Elliptic equations with constant coefficients have associated parabolic problems for which the method of lines approximation admits solutions in terms of matrix exponentials. As described in [17] the common iterative methods for finite difference analogs of elliptic equations can be interpreted as approximations of the exponential matrix. Here we shall consider only the SOR method for standard finite difference approximations to elliptic problems. We shall also restrict ourselves initially to problems with constant coefficients and Dirichlet data on a square. The problem will be imbedded in a parabolic problem, and SOR will be interpreted as a numerical integrator of the method of lines equations. Our approach is thus akin to Garabedian's suggestion where SOR is considered as an integrator for a related hyperbolic equation [5]. We shall summarize and extend the observations given in [16].

Consider the model problem

$$\text{(2.1)}\qquad \begin{aligned} \Delta u &= f, \quad (x,y) \in D = (0,\pi)\times(0,\pi) \\ u &= g, \quad (x,y) \in \partial D, \end{aligned}$$

and the associated parabolic problem

$$\text{(2.2)}\qquad \begin{aligned} 2(2-\omega)v_t &= \omega(\Delta v - f) && (x,y) \in D,\ t > 0 \\ v &= g && (x,y) \in \partial D,\ t > 0 \\ v(x,y,0) &= v_0(x,y) && (x,y) \in D,\ t = 0 \end{aligned}$$

where v_0 is arbitrary. If f is Hölder continuous and g is continuous then it follows immediately that for $\omega \in (0,2)$

$$\lim_{t\to\infty} v(x,y,t) = u(x,y)$$

uniformly in $\bar{D}$ [4, p. 158]. For the parabolic problem (2.2) let us define a uniform mesh on D and use the following non-standard time implicit integrator

$$\text{(2.3)}\qquad \begin{aligned} &4\,\frac{v_{ij}^{n+1} - v_{ij}^{n}}{\Delta t} - \omega\,\frac{v_{i-1j}^{n+1} - v_{i-1j}^{n}}{\Delta t} - \omega\,\frac{v_{ij-1}^{n+1} - v_{ij-1}^{n}}{\Delta t} \\ &= \omega\Big[\frac{v_{i+1j}^{n} + v_{i-1j}^{n} - 2v_{ij}^{n}}{\Delta x^2} + \frac{v_{ij+1}^{n} + v_{ij-1}^{n} - 2v_{ij}^{n}}{\Delta y^2} - f[x_i,y_j)\Big] \end{aligned}$$

where v_{ij}^n is the approximation to $v(x_i,y_j,t_n)$. Since v_{ij}^{n+1} is given on ∂D this time implicit formula is actually explicit in v_{ij}^{n+1} for a standard row-wise ordering of the mesh. It is straightforward to verify that (2.3) is a consistent numerical approximation of (2.2) provided only that $\Delta t \to 0$ implies that $\Delta x \to 0$ and $\Delta y \to 0$. In fact, for the special choice $\Delta t = \Delta x^2 = \Delta y^2$ a Taylor expansion around (x_i,t_i,t_n) of (2.3) shows that the difference equation may be considered, to order Δx^2, an approximation of the hyperbolic equation

$$\text{(2.4)}\qquad \frac{2(2-\omega)}{\omega}\,v_t = v_{xx} + v_{yy} - f - \Delta x(v_{xt} + v_{yt})$$

Moreover, for this choice of discretization we can rewrite (2.3) as

$$-4\tilde{v}_{ij} + v^{n+1}_{i-1j} + v^{n+1}_{ij-1} = -v^{n}_{i+1j} - v^{n}_{ij+1} + \Delta x^2 f(x_i,y_i)$$

(2.5)

$$v^{n+1}_{ij} = v^{n}_{ij} + \omega(\tilde{v}_{ij} - v^{n}_{ij}).$$

Thus, the integrator for the parabolic equation (2.2) is precisely the standard SOR method for the Dirichlet problem (2.1). Alternatively, (2.5) is an integrator for the hyperbolic problem (2.4).

The relationship between (2.4) and (2.5) was exploited in [5] when determining an optimum relaxation factor ω for which the solution of the continuous hyperbolic problem showed fastest decay with time. A method results which does not require any specific structure of the iteration method such as property A and which is applicable whenever the eigenvalues of the Laplacian are known. Here we shall work with (2.3) directly and base the computation of an approximate optimum ω on a Fourier stability analysis.

If we write $w^n_{ij} - v^n_{ij} - u_{ij}$ for the error between the discretized transient and steady state solutions of (2.3) then the error satisfies (2.3) with zero source term and boundary conditions. If $\vec{w}^n$ is the $(N+1)^2$ dimensional vector with components w^n_{ij} (in the rowwise ordering) then $\vec{w}^n \in H_n$ where $\Delta x = \Delta y = \pi/N$, $x_i = i\Delta x$, $y_j = j\Delta y$ and

$$H_N = \operatorname*{span}_{\substack{1\le k\le N-1\\ 1\le \ell\le N-1}} \{\psi_{k\ell}\}$$

$$\psi_{k\ell} = \{\sin kx_i \ \sin \ell y_j\}^N_{i,j=0}$$

An analysis of (2.3) acting on H_N is equivalent to the original SOR analysis and yields no simplification because the basis vectors $\psi_{k\ell}$ are not invariant under the difference formula (2.3). Instead, let us extend $\vec{w}^n$ by periodicity to a regular grid defined on the whole plane and use the basis $\{\psi_{k\ell}\}$ where $\psi_{k\ell}$ has components

$$\psi_{k\ell}(x_i,y_j) = e^{i(kx_i + j\ell y_j)}$$

for all integers $1 < |k| \le N-1$, $1 \le |\ell| \le N-1$. We note that the vectors corresponding to $\bar{k} = \ell = \bar{0}$ and $|\bar{k}| = |\ell| = N$ are excluded because $w^n_{ij} = 0$ on ∂D forces vectors constant in x and y to vanish everywhere. We can now apply the Fourier stability analysis of [14] to the integrator (2.3).

If we write

$$w^n = \sum_{\substack{1 \le |k| \le N-1 \\ 1 \le |\ell| \le N-1}} \alpha^n_{ke} \psi_{k\ell}(x,y)$$

then substitution of w^n into the difference formula (2.3) shows that α^n_{ke} has an associated amplification factor

$$\sigma(k,\ell,\omega,h) = \frac{4(1-\omega) + \omega\bar{z}}{4 - \omega z}$$

where $z = e^{-i\ell\pi h} + e^{-ik\pi h}$ and $h = \Delta t = \Delta x^2 = \Delta y^2$. It is readily found that $|\sigma(k,\ell,0,h)| = |\sigma(k,\ell,2,h)| = 1$ and that $|\sigma(k,\ell,\omega,h)| < 1$ for $\omega \in (0,2)$. Hence the numerical integrator (2.3) is stable for all $\omega \in (0,2)$. We now define our optimum relaxation factor to be the value ω_b which produces the smallest maximum modulus of all the amplification factors for a given mesh h. We observe that

$$\sigma = \frac{y + \bar{z}}{y + (4 - \bar{z})} \qquad \text{where } y = \frac{4}{\omega} - 4$$

so that

$$|\sigma^2| = \frac{y^2 + y\phi + \psi}{y^2 + 8y - y\phi - 4\phi + \psi + 16}$$

for $\phi = z + \bar{z}$ and $\psi = z\bar{z}$. A straightforward calculation shows that

$$\frac{\partial}{\partial\phi}|\sigma^2| > 0 \quad \text{and} \quad \frac{\partial}{\partial\psi}|\sigma^2| > 0.$$

From the definition of ϕ and ψ we see that both are maximized where $k = \ell = 1$. With ϕ and ψ thus fixed the optimum ω can now be computed from

$$\frac{\partial}{\partial y}|\sigma^2| = 0.$$

For the model problem (2.1) we find

$$\omega_b = \frac{2}{1 + 2\sin(\pi/2N)} \tag{2.6}$$

The optimal relaxation factor for the SOR method which minimizes the spectral radius of the iteration matrix is well known to be

$$\tilde{\omega} = \frac{2}{1 + \sin(\pi/N)}$$

so that $|\omega_b - \tilde{\omega}| = O(h^3)$. The equivalent optimum relaxation factor obtained with Garabedian's method is

$$\omega_{b,G} = \frac{2}{1 + \pi h}$$

which approximates $\tilde{\omega}$ to the same order but is not identical with (2.6).

The Fourier method extends, at least in principle, to the fully elliptic problem

$$\text{(2.7)} \qquad \begin{aligned} Lu \equiv au_{xx} + bu_{xy} + cu_{yy} + du_x + eu_y + qu &= f \text{ in } D \\ u &= g \text{ on } \partial D. \end{aligned}$$

The corresponding parabolic problem for $\Delta x = \Delta y = h$ may be chosen as

$$\text{(2.8)} \qquad \begin{aligned} [a(2-\omega) + c(2-\omega) + \frac{\omega d}{2} h + \frac{\omega e}{2} h + h^2 q]v_t &= \omega[Lv-f] \text{ in } D,\ t > 0 \\ v &= g \text{ on } \partial D,\ t > 0 \\ v &= v_0(x,y) \text{ in } D,\ t = 0 \end{aligned}$$

The following time implicit approximation of the left hand side of (2.8)

$$\begin{aligned} &2a \frac{v_{ij}^{n+1} - v_{ij}^{n}}{\Delta t} - a\omega\left(\frac{v_{i-1j}^{n+1} - v_{i-1j}^{n}}{\Delta t}\right) \\ &\quad + \frac{b\omega}{4}\left(\frac{v_{i+1j-1}^{n+1} - v_{i+1j-1}^{n}}{\Delta t} - \frac{v_{i-1j-1}^{n+1} - v_{i-1j-1}^{n}}{\Delta t}\right) \\ &\quad + 2c \frac{v_{ij}^{n+1} - v_{ij}^{n}}{\Delta t} - c\omega\left(\frac{v_{ij-1}^{n+1} - v_{ij-1}^{n}}{\Delta t}\right) \\ &\quad - \frac{\omega dh}{2}\left(\frac{v_{i-1j}^{n+1} - v_{i-1j}^{n}}{\Delta t}\right) - \frac{\omega eh}{2}\left(\frac{v_{ij-1}^{n+1} - v_{ij-1}^{n}}{\Delta t}\right) \\ &\quad + h^2 q \frac{v_{ij}^{n+1} - v_{ij}^{n}}{\Delta t} \end{aligned}$$

and the usual 9 point centered difference approximation Lv-f at time t_n yield the standard SOR difference formula for (2.7) provided again that $\Delta t = \Delta x^2 = \Delta y^2$. An analogous stability analysis now yields the amplification matrix

$$\text{(2.9)} \qquad \sigma(k,\ell,\omega,h) = \frac{A + Be^{ikh} + Ce^{i\ell h} + De^{i\ell h} \cdot \sin kh}{E + Fe^{-ikh} + Ge^{-i\ell h} + De^{-i\ell h} \sin kh}$$

where $A = (2a + 2c - qh^2)(1-\omega)$, $B = (a + \frac{dh}{2})\omega$, $C = (c + \frac{eh}{2})\omega$, $D = \frac{\omega b}{2} i$,

$E = 2a + 2c - h^2 q$, $F = (-a + \frac{dh}{2})\omega$, $G = (-c + \frac{eh}{2})\omega$. The optimum relaxation factor ω_b for (2.8) with periodic rather than Dirichlet boundary conditions is again obtained from the optimization problem

$$\max_{1 \le |h|,|\ell| \le N-1} |\sigma(k,\ell,\omega_b,h)| \le \max_{\substack{1 \le |k|,|\ell| \le N-1 \\ \omega \in (0,2)}} |\sigma(k,\ell,\omega,h)|$$

No attempt has yet been made to exploit the structure of (2.9) to find ω_b. Instead, (2.9) was coded in Fortran complex arithmetic and a search for ω_b was carried out. At the same time an approximate optimum $\tilde{\omega}$ was determined for the discrete Dirichlet problem by counting the iterations required to reduce an initial guess of $w^0_{ij} = 1$ in D and $w^0_{ij} = 0$ on ∂D to $|w^n_{ij}| < 10^{-6}$ uniformly in D. The experiments are summarized in Table 1.

TABLE 1

Comparison of the approximate Fourier relaxation factor ω_b with the experimental best Dirichlet relaxation factor $\tilde{\omega}$ for the problem (2.7) with a = c = d = e = -1 = 1, b = 2. Only $\omega_i = 2i/20$, i = 0,...,20 were examined. The optimum $\tilde{\omega}$ for N = 100 proved too expensive to compute.

N	ω_b	$\tilde{\omega}$
10	1.45	1.50
20	1.60	1.60
40	1.70	1.75
100	1.80	*

The above approach for determining ω_b may be of particular value when coupled with the ad-hoc SOR method suggested in [2] where variable coefficient equations are relaxed with a local relaxation factor determined by assuming locally constant coefficient difference equations. Since the above method applies to the full elliptic equation (2.7) more general domains can be mapped to squares for which positive approximate relaxation factors can be obtained without spectral information on the iteration matrix. Moreover, if the maximum absolute relaxation factor can be shown to lie on boundary of the admissible Fourier frequencies (see, e.g., the derivation of (2.6)) then ω_b can be found with miminal work regardless of the size of the matrix system.

3. The Method of Lines for Elliptic Problems. When all but one of the independent variables of an elliptic boundary value problem are discretized, a boundary value problem for a system of ordinary differential equations results. A description of the classical method of lines and some comments on its implementation may be found in [11], [7]. A rationale for using the method of lines and a convergence analysis for the method of moments based on global rather than local discretizations in directions orthogonal to the continuous variable may be found in [9], [10], and a typical application of the method of lines to fracture mechanics is described in [6]. However, as a general purpose routine the method of lines is generally held to be inferior to finite difference and finite element solvers for standard elliptic boundary value problems.

In our work the method of lines is used primarily for non-standard problems such as non-linear problems on irregular domains with free boundaries. In this setting we find the method useful because the intersection of each line with the boundary is readily visualized. This simplifies the interpolation of data on adjacent lines near the boundary which is usually required to compute along a given line. On the other hand, this setting largely precludes the use of special techniques such as fundamental matrices or complementary functions when solving the boundary value problem for the approximating system of ordinary differential equations. Instead, we shall solve the system iteratively one at a time with what is essentially a line SOR method. A sequentially one-dimensional algorithm results with considerable flexibility in treating unusual boundary conditions. Overall, we judge from our numerical experiments that this simple locally one-dimensional algorithm produces answers which are comparable in accuracy to those obtained with domain mapping methods where the physical domain is transformed to a simple computational domain [15].

Let us illustrate this use of the method of lines. A well studied elliptic free boundary problem models the steady state seepage of water through an earthen dam. Assuming a homogeneous dam and the geometry shown in Figure 1 we obtain the following equations and boundary conditions for the piezometric head $y = p + y$.

$$\Delta u = 0 \text{ in } D$$

$$u = h_1 \text{ on the left face of the dam}$$

(3.1)
$$u = \max\{h_2, y\} \text{ on the right face of the dam}$$

$$\frac{\partial u}{\partial n} = 0 \text{ across the base of the dam.}$$

Here D is the saturated region bounded above by the free boundary ∂D which is determined from the Cauchy conditions

(3.2)
$$u = y; \qquad \frac{\partial n}{\partial u} = 0.$$

For a sloping rear face a solution u is acceptable when $\partial u/\partial n \leq 0$ below the free boundary which assures that water will seep out below the seepage point.

It is possible, and often recommended in the literature, to transform the original free boundary problem to a fixed domain problem with the so-called Landau transformation $\eta = \frac{y}{s(x)}$. While the computational domain becomes a rectangle and the method of lines approixmation a two-point boundary value problem, the differential equations are somewhat complicated to derive and solve (see, e.g., [15]). The line SOR approach advocated below appears to have comparable accuracy, works under same conditions as the Landau transformation but is easier to formulate and implement since the original equations are used.

Problem (3.1) is characteristic of the type of problems for which the method of lines is sgugested. We have an irregular domain and non-standard boundary conditions. On the other hand, the free boundary is required to be smooth and single valued. In earlier work on a square dam with variable permeabilities the seepage point was difficult to find with the method of lines when x was discretized because the free boundary became tangent to the line of computation near the back face [13]. As suggested there polar coordinates should be used for the method of lines discretization. If θ_0 and θ_M denote the angles of the base and front face and if

$$\theta_i = \theta_0 + \frac{i(\theta_M - \theta_0)}{M}, \qquad i = 0,\ldots,M$$

denote M+1 evenly spaced rays then along the ray $\theta = \theta_i$ the method of lines based on central finite difference approximations in θ yields the following approximation of (3.1)

$$L_i u_i = (ru_i')' - \frac{2}{r\Delta\theta^2} u_i = -\frac{1}{r\Delta\theta^2}(u_{i+1} + u_{i-1}), \tag{3.3}$$

$$i = 0,\ldots,M-1$$

The boundary conditions on the base and front face imply

$$u_i(0) = h,$$

$$u_M(r) = h,$$

$$u(r) = u_1(r).$$

On the free boundary ∂D, expressed as $r = s(\theta)$ the boundary conditions are rewritten as

$$u(r,\theta) = s(\theta)\sin\theta$$

$$\frac{\partial u}{\partial r} = \frac{s'(\theta)^2 \sin\theta + s(\theta)s'(\theta)\cos\theta}{s(\theta)^2 + s'(\theta)^2}$$

and discretized as

$$u_i(s_i) = s_i \sin\theta_i$$

$$u_i'(s_i) = \frac{((s_{i+1}-s_{i-1})/2\Delta\theta)^2 \sin\theta_i + s_i(s_{i+1}-s_{i-1})2\Delta\theta \cos\theta_i}{\left(\frac{s_{i+1}s_{i-1}}{2\Delta\theta}\right)^2 + s_i^2}$$

If no free boundary is found along the ray $\theta = \theta_i$ we set $u_i(s_i) = \max\{h_2, s_i \sin\theta_i\}$ for s_i on the rear face. Thus, the system (3.1,2) is replaced by a multi-point free boundary problem for ordinary differential equations. The sequentially one dimensional method may be described as follows. Given an initial guess $\{u_i^0\}$ and $\{s_i^0\}$ for $i = 0,\ldots,M-1$ we compute for $i = 0,\ldots,M-1$ and $k = 1,2,\ldots$ an intermediate solution $\{\tilde{u}_i, s_i^k\}$ from the scalar free boundary problem

$$L\tilde{u}_i = F_i^k(r)$$

$$\tilde{u}_i(0) = h_1, \tilde{u}_i(s_i^k) = s_i^k \sin\theta_i \tag{3.4}$$

$$\tilde{u}_i'(s_i^k) = g_i^k(s_i^k)$$

with

$$F_i^k(r) = \frac{1}{r\Delta\theta^2}\left[u_{i-1}^k(r) + u_{i+1}^{k-1}(r)\right]$$

$$g_i^k(r) = \frac{s_i^{k'^2} \sin\theta_i + r s_i^{k'} \cos\theta_i}{s_i^{k'^2} + r^2}$$

where

$$s_i^{k'} = \frac{s_{i+1}^{k-1} - s_{i-1}^k}{2\Delta\theta}$$

The numerical solution of (3.4) is relatively simple because a single second order equation must be solved. In our work we find the sweep method (invariant imbedding) convenient because it uncouples the free boundary from the solution of the differential equations.

The sweep method for free boundary problems has been described in detail before and will only be outlined here. The equation (3.4) is written as a first order system

$$r\tilde{u}_i' = v_i$$

(3.5)

$$v_i' = \frac{2}{r\Delta\theta^2}\tilde{u}_i + F_i^k(r).$$

Then $\tilde{u}_i$ and v_i are related through the Riccati transformation

$$\tilde{u}_i = R(r)v_i + w_i^k(r) \tag{3.6}$$

where R and w_i^k are the solutions of the initial value problem

$$R' = \frac{1}{r} - \frac{2}{r\Delta\theta^2}R^2 \qquad R(0) = 0 \tag{3.7}$$

$$w_i^{k'} = -\frac{2}{r\Delta\theta^2}R(r)w_i^k(r) - R(r)F_i^k(r) \qquad w_i^k(0) = h_1. \tag{3.8}$$

The solution R is known in closed form but is obtained here numerically from a perturbed initial point $0 < r_0 << 1$ so that the resulting computer code is not tied to Laplace's equation. Similarly, the integration w_i^k is started at r_0. Both equations are solved with a fourth order Adams-Moulton method defined on a fixed mesh along each ray. Since (3.7) and (3.8) are quadratic and linear the implicit difference equations at each mesh point can be solved in closed form. The use of a variable order Adams-Moulton method would be preferable for computational efficiency. A variable mesh method, on the other hand, would require extensive interpolation to communicate data between adjacent rays through the source term $F_i^k(r)$. With R(r) and $w_i^k(r)$ available the free boundary is found by searching for the value s_i^k where $\tilde{u}_i(r)$ and $v_i(r)$ satisfies

$$\tilde{u}_i(r) = R(r)\frac{\tilde{u}_i'(r)}{r} + w_i^k(r)$$

$$\tilde{u}_i(r) = r \sin\theta_i$$

$$\tilde{u}_i'(r) = g_i^k(r).$$

Elimination of $\tilde{u}_i$ and $\tilde{u}_i'$ from these three expressions shows that s_i^k

must be a root of the scalar function

$$\phi_i^k(r) \equiv r \sin\theta_i - R(r)\frac{g_i^k(r)}{r} - w_i^k(r) = 0.$$

The first positive root of $\phi_i^k(r) = 0$ is chosen as s_i^k. If no root is found within the dam along the ray $\theta = \theta_i$ then s_i^k is the intersection between the ray and the rear face of the dam.

A numerical solution of the problem for a square dam in cartesian coordinates with continuous y-coordinate is analyzed in [13]. While convergence of a line Gauss-Seidel method can be established the numerical solution deteriorated near the seepage point because the free surface becomes tangent to the line of computation. Polar coordinates eliminated all problems and gave answers which remained stable with respect to changes in mesh sizes. Figure 1 shows a plot of the computed drainage curve and, for comparison, the finite element solution and the analytic solution of [3].

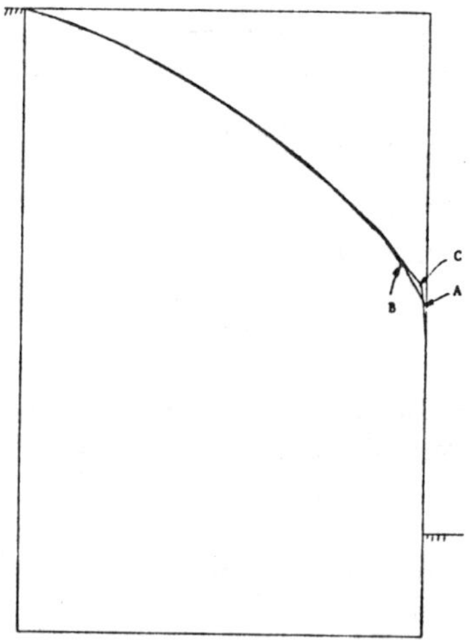

Figure 1. Linearly interpolated drainage curves for the model problem of [3]. 20 rays, 150 mesh points per ray. A: exact, B: finite element; C: MOL solution.

The question remains whether this approach to two point boundary value problems presents anything new. In our view the difference lies in how the problem is interpreted with the sweep method rather than in the final algorithm. Indeed, it is well known that the sweep method for the solution of the differential equations (3.7,8) along a given line may be considered the limiting case as $\Delta r \to 0$ of Gaussian elimination applied to a discretized form of (3.4) (see, e.g., [1], [8]). This relationship can actually be made exact if discrete invariant imbedding is applied.

Consider a uniform mesh $r_0 < r_1 < \dots < r_N$ along each ray where r_N is chosen large enough to be beyond the free boundary. Let us discretize (3.5) with the following non-symmetric implicit/explicit Euler method

$$\frac{\tilde{u}_{j+1} - \tilde{u}_j}{\Delta r} = \frac{v_{j+1}}{r_{j+\frac{1}{2}}}$$

(3.9)

$$\frac{v_{j+1} - v_j}{\Delta r} = \frac{2}{r_j \Delta\theta^2}\,\tilde{u}_j + F_i^k(r_j)$$

where j denotes the jth mesh point along the ray $\theta = \theta_i$ and where the subscript i is suppressed except in the source term. If $\{\tilde{u}_j, v_j\}_{j=1}^{N-1}$ are solutions for those two difference equations then it follows from substituting the first into the second equation that $\{\tilde{u}_j\}$ satisfies

$$\frac{1}{\Delta r}\left[r_{j+\frac{1}{2}}\frac{\tilde{u}_{j+1}-\tilde{u}_j}{\Delta r} - r_{j-\frac{1}{2}}\frac{\tilde{u}_j-\tilde{u}_{j-1}}{\Delta r}\right] - \frac{2}{r_j\Delta\theta^2}\,\tilde{u}_j = F_i^k(r_j).$$

Hence the solution of (3.9) is exactly the same as the line SOR solution along the ray $\theta = \theta_i$ obtained from a standard central difference analog applied to (3.4). If we now apply discrete invariant imbedding to (3.9) then (3.6) is replaced by

(3.10) $$\tilde{u}_j = R_j v_j \; y \; w_j$$

where R_j and w_j are found to satisfy the difference equations (see [12, p. 29])

(3.11) $$\frac{R_{j+1}-R_j}{\Delta r} = \frac{1}{r_{j+\frac{1}{2}}} - \frac{2}{r_j\Delta\theta^2} R_j R_{j+1} + \frac{2r}{r_j\Delta\theta^2}\,\frac{R_j}{r_{j+\frac{1}{2}}}\,, \quad R_0 = 0$$

$$\frac{w_{j+1}-w_j}{\Delta r} = -\frac{2R_{j+1}}{r_j\Delta\theta^2} w_j - R_{j+1}F_i^k(r_j)$$

$$+\left(\frac{2}{r_j\Delta\theta^2} w_j - F_i^k(r_j)\right)\frac{\Delta r}{r_{j+\frac{1}{2}}}, \quad w_0 = h_1.$$

It is readily seen that the equations (3.11) define a consistent numerical integrator for the sweep equations (3.7,8). In summary, we have shown that for a special choice of numerical integrator for the invariant imbedding equations the discrete solution of the sweep method is exactly the same as the discrete solution of the standard line SOR method applied to the central difference analog of the elliptic partial differential equation.

So far no systematic comparison of various integrators for the sweep method has been carried out. The choice of an Adams code for (3.7,8) was based solely on convenience.

Finally, we would like to point out that the sequentially one-dimensional approach to multi-dimensional free boundary problems is not restricted to single equations in several variables but may also be used for systems of equations. For example, one may apply the above algorithm to the reaction-diffusion system

(3.12a) $$\frac{\partial T}{\partial t} = D_1\Delta T + Qn\ \exp(C/T)$$

(3.12b) $$\frac{\partial n}{\partial t} = D_2\Delta n - n\ \exp(C/T)$$

defined on the rectangle $(0,\bar{X}) \times (0,\bar{Y})$. We shall assume Dirichlet data on $y = 0$, Neumann data on $x = 0$ and $x = \bar{X}$ and, somewhat arbitrarily, Neumann data on n at $y = \bar{Y}$ and a free boundary condition like

(3.13) $$u = u_0$$
$$\frac{\partial u}{\partial n} = \varepsilon$$

on the free boundary $y = s(x,t)$. Alternatively, n may have its own free boundary.

The line SOR method for the system (3.12) can be based on the following linearization. In iteration k along line i we approximate the Arrhenius term by the first two terms of the Taylor expansion

$$\left(e^{C/T}\right)_i^k = e^{\frac{C}{T_i^{k-1}}} - \frac{C}{{T_i^{k-1}}^2} e^{\frac{C}{T_i^{k-1}}} (T_i^k - T_i^{k-1})$$

and apply the above algorithm to this single equation in T_i^k for given n. Once (T_i^k, s_i^k) have been found we solve for n and its free boundary and then proceed to the next line. Initial experiments with representative reaction rates produce convergence; however, the full power of monotonicity arguments remains to be applied to suitably rewrite equations (3.12) and assure a priori convergence. This work is continuing.

REFERENCES

[1] I. BABUSKA, The connection between finite difference like methods based on initial value problems for ODE, in Numerical Solutions of Boundary Value Problems for Ordinary Differential Equations, A. K. Aziz, edt., Academic Press, N.Y., 1975.

[2] L. W. EHRLICH, An ad-hoc SOR method, J. Comp. Phys. 44 (1981), pp. 31-45.

[3] C. M. ELLIOTT and J. R. OCKENDON, Weak and Variational Methods for Moving Boundary Problems, Research Notes in Mathematics, No. 59, Pitman, London, 1982.

[4] A. FRIEDMAN, Partial Differential Equations of Parabolic Type, Prentice-Hall, Englewood Cliffs, N.J., 1964.

[5] P. R. GARABEDIAN, Estimation of the relaxation factor for small mesh size, Math. Tables Aids Comput. 10 (1956), pp. 183-185.

[6] J. P. GYEKENYESI and A. MENDELSON, Three-dimensional elastic stress and displacement analysis of finite geometry solids containing cracks, Int. J. Fracture 11 (1975), pp. 409-429.

[7] D. J. JONES and J. C. SOUTH, JR., Application of the method of lines to the solution of elliptic partial differential equations, NRC No. 18021, Aeronautical Report LR-599, National Research Council Canada, Ottawa, 1979.

[8] H. B. KELLER and M. LENTINI, Invariant imbedding, the box scheme and an equivalence between them, SIAM J. Num. Anal. 19 (1982), pp. 942-962.

[9] S. H. LEVENTHAL, Method of moments for singular problems, Comp. Methods Appl. Mech. Engng. 6 (1975), pp. 79-100.

[10] S. H. LEVENTHAL, The method of moments and its optimization, Int. J. Num. Methods Engng. 9 (1975), pp. 337-351.

[11] O. A. LISKOVETS, The method of lines, Differential Equations 1 (1965), pp. 1308-1323.

[12] G. M. MEYER, Initial Value Methods for Boundary Value Problems, Academic Press, New York, 1973.

[13] G. H. MEYER, On the computational solution of elliptic and parabolic free boundary problems, in Free Boundary Problems, E. Magenes, edt., Istituto Nazionale di Alta Matematica Francesco Severi, Rome, 1980.

[14] R. D. RICHTMYER nad K. W. MORTON, Difference Methods for Initial-Value Problems, Interscience, New York, 1967.

[15] T. SAITOH, Numerical method for multidimensional freezing problems in arbitrary domains, Journal Heat Man. Transfer 100 (1978), pp. 294-299.

[16] K. J. SCHWENZFEGER, edt., SEMINAR ÜBER DIE LINIENMETHODE, Hochschule der Bundeswehr, Neubiberg, Germany, 1982.

[17] R. S. VARGA, Matrix Iterative Analysis, Prentice Hall, Englewood Cliffs, N.J., 1962.

PART II: THE MATHEMATICS OF COMBUSTION AND CHEMICAL REACTIONS

DETERMINATION OF A RATE CONSTANT AND AN ACTIVATION ENERGY IN SOME EXOTHERMIC CHEMICAL REACTIONS

JOHN R. CANNON,* JAMES C. CAVENDISH** AND JOSEPH V. KOEBBE†

Abstract. A very old problem of chemical kinetics is that of estimating the rate constant, k, and the activation energy, E, in a chemically reacting system. In this paper we consider use of dynamic temperature measurements for determining k and E in a simple first order exothermic chemical reaction. First, we develop proofs which demonstrate the continuous dependence of k and E upon given physical data (initial conditions and temperature, for example) from which follows uniqueness in the determination of k and E. Next, the non-linear least-squares method is used to fit these kinetic parameters to temperature data measured over a significant portion of the transient period. A priori and a posteriori error estimates are derived for the least-squares estimates, and the results of numerical calculations are used to study the effects on parameter estimation arising from data measurement errors (both random and systematic) which invariably accompany measurement of the output of any chemical process.

*Mathematics Department, Washington State University, Pullman, WA 99164-2930

**Mathematics Department, General Motors Research Laboratories, Warren, Michigan 48090

†Mathematics Department, University of Wyoming, Laramie, Wyoming 82071

1. Introduction. The general problem of estimation of parameters in differential equations arises in many fields of engineering and science. An important example of this is the estimation of rate constants (sometimes called preexponential factors) and activation energies in exothermic chemical reactions. To take a simple case, suppose we consider the exothermic reaction system

$$A \xrightarrow{k} B + \text{Heat} \tag{1.1}$$

and assume that it is described kinetically by the first order differential system

$$\begin{aligned} \dot{A} &= \frac{dA}{dt} = -k \exp(-E/RT)\, A \\ \dot{B} &= \frac{dB}{dt} = k \exp(-E/RT)\, A \\ \dot{T} &= \frac{dT}{dt} = \frac{q}{c}\, k \exp(-E/RT)\, A \\ A(0) &= A_0 \\ B(0) &= B_0 \\ T(0) &= T_0 \end{aligned} \tag{1.2}$$

where the initial concentrations A_0 and B_0 of the chemical species A and B are nonnegative, the initial temperature T_0 is positive, and q, R and c represent, respectively, the heat of reaction, the universal gas constant and the system heat capacity at constant volume. The rate of reaction, k, and the activation energy, E, are positive constants, and the dot denotes differentiation with respect to the time variable, t. A very old problem of chemical kinetics is that of estimating the rate constant, k and the activation energy, E. We assume that the actual dynamics of the reacting system is governed by equations (1.1), (1.2) with given initial data A_0^*, B_0^*, T_0^*, and system constants q^*, R^* and c^*, and we denote by $T^*(t)$ the corresponding transient temperature component of the solution. Mathematically, we pose the problem of determining positive unknown constants, k and E, given the data A_0^*, B_0^*, T_0^*, q^*, R^*, c^* and $T^*(t)$ where $T^*(t)$ is specified over the time interval $0 \le t_1 \le t \le t_2$.

Using the fact that $\dot{T} + \frac{q}{c}\dot{A} = 0$, which arises from simple manipulations of equation (1.2), we see that T satisfies the following initial value problem:

$$\dot{T} = k\,\exp(-E/RT)\,(T_0 + \tfrac{q}{c}A_0 - T)$$

(1.3) $$T(0) = T_0$$

It follows from the theory of solutions of ordinary differential equations [4] that the solution

(1.4) $$T = T(t,\ A_0,\ T_0,\ q,\ c,\ R,\ k,\ E)$$

depends continuously upon its arguments t, A_0, T_0, q, c, r, k and E.

<u>Definition</u>: A <u>solution</u> of the mathematical problem is a pair (κ,ξ) such that for $t_1 \le t \le t_2$, $T(t,\ A_0^*,\ T_0^*,\ q^*,\ c^*,\ R^*,\ \kappa,\ \xi) \equiv T^*(t)$.

In Section 2 we will show that the solution depends continuously upon the data A_0^*, T_0^*, q^*, c^*, R^*, and T^*. Uniqueness of the solution will then follow as a corollary to this continuous dependence result. In practice, it is possible to measure system initial data and parameters only approximately. Therefore, in Section 3 it will be assumed that A_0^*, T_0^*, q^*, c^*, R^* and T^* are known only approximately (say, because of experimental measurement error) as $\tilde{A}_0^*$, $\tilde{T}_0^*$, $\tilde{q}^*$, $\tilde{c}^*$, $\tilde{R}^*$ and $\tilde{T}^*$ such that

$$\max\{|A_0^*-\tilde{A}_0^*|,\ |T_0^*-\tilde{T}_0^*|,\ |q^*-\tilde{q}^*|,\ |c^*-\tilde{c}^*|,\ |R^*-\tilde{R}^*|,\ \|T^*-\tilde{T}^*\|\} < \varepsilon ,$$

where ε is an estimate (known or unknown) of the error in the measured data, and where

$$\|T^*-\tilde{T}^*\| = \sup_{t_1 \le t \le t_2} |T^*(t) - \tilde{T}^*(t)| .$$

The method of nonlinear least-squares will be presented for the numerical approximation of the solution. A priori and a posteriori error estimates will be summarized and the reader will be referred to Cannon and Filmer [1] for proofs. A generalization of the problem for the system of equations

(1.6)
$$\dot{\vec{y}} = \vec{F}(t,\ \vec{y},\ \vec{k})$$
$$\vec{y}(0) = \vec{y}_0$$

is presented in Section 4 and this analysis corrects the analysis presented in [1] for a similar generalization. The paper is concluded with a discussion of results of some numerical experiments in Section 5 and another example reaction in Section 6.

2. Continuous Dependence Upon the Data. Suppose that $(\kappa^{(i)}, \xi^{(i)})$, $i = 1,2$, are solutions which correspond respectively to the data $A_0^{(i)*}$, $T_0^{(i)*}$, $q^{(i)*}$, $c^{(i)*}$, $R^{(i)*}$ and T_i^* $i = 1,2$. From the differential system (1.2), we have that

$$\dot{T}_i(0) = \kappa^{(i)} \frac{q^{(i)*}}{c^{(i)*}} A_0^{(i)*} \exp(-\xi^{(i)}/R^{(i)*}T_0^{(i)*}) \tag{2.1}$$

and

$$\frac{\ddot{T}_i(0)}{(\dot{T}_i(0))^2} = \frac{\xi^{(i)}}{R^{(i)*}(T_0^{(i)*})^2} - \frac{1}{\frac{q^{(i)*}}{c^{(i)*}} A_0^{(i)*}} . \tag{2.2}$$

We consider the differences $\dot{T}_2(0) - \dot{T}_1(0)$ and $\dfrac{\ddot{T}_2(0)}{(\dot{T}_2(0))^2} - \dfrac{\ddot{T}_1(0)}{(\dot{T}_1(0))^2}$.

First,

$$\dot{T}_2(0) - \dot{T}_1(0) = \int_0^1 \frac{d}{d\theta} F(\theta)\, d\theta , \tag{2.3}$$

where

$$F(\theta) = \bar{K}(\theta) \frac{\bar{q}(\theta)}{\bar{c}(\theta)} \exp(-\bar{E}(\theta)/\bar{R}(\theta)\bar{T}_0(\theta))\bar{A}_0(\theta) \tag{2.4}$$

and

$$\begin{aligned}
\bar{K}(\theta) &= \kappa^{(1)}(1-\theta) + \kappa^{(2)}\theta \\
\bar{q}(\theta) &= q^{(1)*}(1-\theta) + q^{(2)*}\theta \\
\bar{c}(\theta) &= c^{(1)*}(1-\theta) + c^{(2)*}\theta \\
\bar{E}(\theta) &= \xi^{(1)}(1-\theta) + \xi^{(2)}\theta \\
\bar{R}(\theta) &= R^{(1)*}(1-\theta) + R^{(2)*}\theta \\
\bar{T}_0(\theta) &= T_0^{(1)*}(1-\theta) + T_0^{(2)*}\theta \\
\bar{A}_0(\theta) &= A_0^{(1)*}(1-\theta) + A_0^{(2)*}\theta
\end{aligned} \tag{2.5}$$

and

$$\frac{dF}{d\theta} = F(\theta)\left(\frac{\kappa^{(2)}-\kappa^{(1)}}{\overline{K}(\theta)} + \frac{q^{(2)*}-q^{(1)*}}{\overline{q}(\theta)} + \frac{A_0^{(2)*}-A_0^{(1)*}}{\overline{A}_0(\theta)} - \frac{c^{(2)*}-c^{(1)*}}{\overline{c}(\theta)^2}\right.$$

$$\left. - \frac{\xi^{(2)}-\xi^{(1)}}{\overline{R}(\theta)\overline{T}_0(\theta)} + \frac{(R^{(2)*}-R^{(1)*})E(\theta)}{\overline{R}(\theta)^2\overline{T}_0(\theta)} + \frac{(T_0^{(2)*}-T_0^{(1)*})E(\theta)}{\overline{R}(\theta)\overline{T}_0(\theta)^2}\right). \tag{2.6}$$

Secondly,

$$\frac{\ddot{T}_2(0)}{(\dot{T}_2(0))^2} - \frac{\ddot{T}_1(0)}{(\dot{T}_1(0))^2} = \int_0^1 \frac{d}{d\theta}G(\theta)d\theta ,$$

where

$$G(\theta) = \frac{\overline{E}(\theta)}{\overline{R}(\theta)\overline{T}_0(\theta)^2} - \frac{\overline{c}(\theta)}{\overline{q}(\theta)\overline{A}_0(\theta)} , \tag{2.7}$$

$$\frac{dG}{d\theta} = \frac{\xi^{(2)}-\xi^{(1)}}{\overline{R}(\theta)\overline{T}_0(\theta)^2} - \frac{(R^{(2)*}-R^{(1)*})\overline{E}(\theta)}{\overline{R}(\theta)^2\overline{T}_0(\theta)^2} - \frac{2(T_0^{(2)*}-T_0^{(1)*})\overline{E}(\theta)}{\overline{R}(\theta)\overline{T}_0(\theta)^3}$$

$$- \frac{(c^{(2)*}-c^{(1)*})-(q^{(2)*}-q^{(1)*})\overline{c}(\theta)\overline{q}(\theta)^{-1}-(A_0^{(2)*}-A_0^{(1)*})\overline{c}(\theta)\overline{A}_0(\theta)^{-1}}{\overline{q}(\theta)\overline{A}_0(\theta)} \tag{2.8}$$

From (2.3) through (2.8) we obtain the system of equations

$$\begin{aligned} \alpha(\kappa^{(2)}-\kappa^{(1)}) + \beta(\xi^{(2)}-\xi^{(1)}) &= D_1 , \\ \gamma(\xi^{(2)}-\xi^{(1)}) &= D_2 , \end{aligned} \tag{2.9}$$

where

$$\alpha = \int_0^1 \frac{F(\theta)}{\overline{K}(\theta)}d\theta \tag{2.10}$$

$$\beta = \int_0^1 \frac{-F(\theta)}{\overline{R}(\theta)\overline{T}_0(\theta)}d\theta \tag{2.11}$$

$$(2.12) \qquad \gamma = \int_0^1 \frac{1}{\overline{R}(\theta)\overline{T}_0(\theta)^2}\, d\theta \; .$$

D_1 is a linear combination of $(\dot{T}_2(0)-\dot{T}_1(0))$ and the quantities $(q^{(2)*}-q^{(1)*})$, $(A_0^{(2)*}-A_0^{(1)*})$, $(c^{(2)*}-c^{(1)*})$, $(R^{(2)*}-R^{(1)*})$, and $(T_0^{(2)*}-T_0^{(1)*})$, and D_2 is a linear combination of $(\ddot{T}_2(0)/\dot{T}_{(2)}(0)^2) - (\ddot{T}_1(0)/\dot{T}_1(0)^2)$, and the quantities in D_1 except $(\dot{T}_2(0)-\dot{T}_1(0))$.

In order to estimate the solution of the system (2.9) we must obtain lower bounds on α and γ and a upper bound on β. At this point it is convenient to define the set

$$(2.13) \qquad H = \{(A_0,\, T_0,\, q,\, c,\, R,\, \kappa,\, \xi)\}$$

such that

$$A_{0*} \le A_0 \le A_0^{**}\, , \quad T_{0*} \le T_0 \le T_0^{**}$$
$$q_* \le q \le q^{**}\, , \quad c_* \le c \le c^{**}\, ,$$
$$R_* \le R \le R^{**}\, , \quad \kappa_* \le \kappa \le \kappa^{**}$$
$$\xi_* \le \xi \le \xi^{**}\, ,$$

where A_{0*}, A_0^{**}, T_0^*, T_0^{**}, q_*, q^{**}, c_*, c^{**}, R_*, R^{**}, κ_*, κ^{**}, ξ_*, and ξ^{**} are known positive constants. We assume that the data $A_0^{(i)*}$, $T_0^{(i)*}$, $q^{(i)*}$, $c^{(i)*}$ and $R^{(i)*}$, $i=1,2$ belong to the set H. Then from (2.4) and (2.10) we have that

$$(2.14) \qquad \alpha \ge \frac{q_* A_{0**}}{c^{**}} \exp(-\xi^{**}/R_* T_{0*}) > 0\, ,$$

while from (2.11)

$$(2.15) \qquad |\beta| \le \frac{q^{**} A_0^{**} \kappa^{**}}{c_* R_* T_{0*}}$$

and from (2.12)

$$(2.16) \qquad \gamma \ge \frac{1}{R_* T_{0*}^2} > 0\, .$$

Estimates similar to those for β are easily obtained for the coefficients of the quantities $(q^{(2)*}-q^{(1)*})$, $(A_0^{(2)*}-A_0^{(1)*})$, $(c^{(2)*}-c^{(1)*})$, $(R^{(2)*}-R^{(1)*})$, and $(T_0^{(2)*}-T_0^{(1)*})$ which appear in D_1 and D_2. Consequently, we need only estimate $(\dot{T}_1(0)-\dot{T}_1(0))$ and $(\ddot{T}_2(0)/\dot{T}_2(0)^2)-(\ddot{T}_1(0)/\dot{T}_1(0)^2)$ in terms of the data $T_2^*-T_1^*$.

Since the term

$$(\ddot{T}_2(0)/\dot{T}_2(0)^2) - (\ddot{T}_1(0)/\dot{T}_1(0)^2) =$$

$$\dot{T}_2(0)^{-2}\dot{T}_1(0)^{-2}(\dot{T}_1(0)^2(\ddot{T}_2(0)-\ddot{T}_1(0)) + \ddot{T}_1(0)(\dot{T}_1(0)^2-\dot{T}_2(0)^2)), \tag{2.17}$$

we derive upper and lower bounds on $\dot{T}_1$, $i=1,2$ and an upper bound on $\ddot{T}_1(0)$. We begin with the estimates of $\dot{T}_1(0)$. From (2.1), it follows that

$$\dot{T}_1(0) \ge \frac{\kappa_* q_* A_{0*}}{c^{**}} \exp(-\xi^{**}/R_* T_{0*}) \tag{2.18}$$

while

$$\dot{T}_1(0) \le \frac{\kappa^{**} q^{**} A_0^{**}}{c^*}. \tag{2.19}$$

Using (2.19) with (2.2), we see that

$$|\ddot{T}_1(0)| \le \left(\frac{\kappa^{**} q^{**} A_0^{**}}{c_*}\right)^2 \left(\frac{\xi^{**}}{R_* T_{0*}^2} + \frac{c^{**}}{q_* A_{0*}}\right) \tag{2.20}$$

Hence, we need to estimate $(\dot{T}_2(0)-\dot{T}_1(0))$ and $(\ddot{T}_2(0)-\ddot{T}_1(0))$ in terms of $(T_2^*-T_1^*)$ in order to complete the derivation of continuous dependence upon the data.

To that end, let

$$\Omega = \{T(t,A_0,T_0,q,c,R,\kappa,\xi) \mid (A_0,T_0,q,c,R,\kappa,\xi) \in H\}. \tag{2.21}$$

By replacing t by the complex variable $\zeta = t+i\tau$ in the solution T of (1.3), Ω becomes a family of analytic functions. From the proof of the existence of the solution of (1.3), there exists $t_* < 0$ and a $K > 0$ such that

(2.22) $$|T| \le K$$

for every $T \in \Omega$ and for $(t,\tau) \in S$ where

(2.23) $$S = \{(t,\tau) \mid t_* \le t \le t_2,\ -\delta \le \tau \le \delta\}$$

and δ is sufficiently small and fixed. Hence, there exists a positive constant K such that for all ζ in S,

(2.24) $$\left|\frac{d^i T}{dt^i}\right| \le K\ , \quad i = 1,2,3$$

for all T in Ω.

Lemma 1: If $(A_0^{(i)*}, T_0^{(i)*}, q^{(i)*}, c^{(i)*}, R^{(i)*}, \kappa^{(i)}, \xi^{(i)}), i=1,2,$ belong to H, then there exist positive constants K and ω, $0 < \omega < 1$, which depend on H and S, such that

(2.25) $$|\dot{T}_2(0)-\dot{T}_1(0)| + |\ddot{T}_2(0)-\ddot{T}_1(0)| \le K\|T_2^*-T_1^*\|^{\omega}\ .$$

Proof: The proof of this lemma is similar to that given in [1].

We can now state our main result of this section:

Theorem 1: If $(A_0^{(i)*}, T_0^{(i)*}, q^{(i)*}, c^{(i)*}, R^{(i)*}, \kappa^{(i)}, \xi^{(i)})$, $i = 1,2$, belong to H, then there exist constants K and ω, $0 \le \omega \le 1$, which depend on H and S, such that

$$|\kappa^{(2)}-\kappa^{(1)}| + |\xi^{(2)}-\xi^{(1)}| \le$$

$$K(\|T_2^*-T_1^*\|^{\omega} + |q^{(2)*}-q^{(1)*}| + |A_0^{(2)*}-A_0^{(1)*}|$$

(2.26) $$+ |c^{(2)*}-c^{(1)*}| + |R^{(2)*}-R^{(1)*}| + |T_0^{(2)*}-T_0^{(1)*}|)$$

Proof: Apply the Lemma and the estimates (2.14) through (2.20) to the system (2.9).

As a corollary to the Theorem, we obtain the following uniqueness result:

Corollary 1: For each set of data A_0, T_0, q, c, R, and T, there can exist at most one solution (κ,ξ) for which $(A_0, T_0, q, c, R, \kappa, \xi)$ is in H.

In addition, we can state an additional result:

Corollary 2: *In the statement (2.26), the $\|T_2^*-T_1^*\|^{\omega}$ can be replaced with $\|T_2^*-T_1^*\|_2^{\omega}$ for a suitable change of K, where*

$$\|T_2^*-T_1^*\|_2 = \left(\int_{t_1}^{t_2} (T_2^*(t)-T_1^*(t))^2 dt\right)^{1/2} \tag{2.27}$$

Proof: See the corresponding corollary and its proof in [1].

3. A Nonlinear Least-squares Approximation Method and Estimates of Error. We assume that $(A_0^*, T_0^*, q^*, c^*, R^*, \kappa, \xi)$ and $(\tilde{A}_0^*, \tilde{T}_0^*, \tilde{q}^*, \tilde{c}^*, \tilde{R}^*, \kappa, \xi)$ are in H. Next, let

$$H^* = \{(\kappa, \xi) \mid \kappa_* \le \kappa \le \kappa^{**},\ \xi_* \le \xi \le \xi^{**}\} . \tag{3.1}$$

Also, set

$$\tau_j = t_1 + \frac{j(t_2-t_1)}{n} , \quad j = 0,1,2,\ldots,n , \tag{3.2}$$

and define

$$\tilde{T}(t,\kappa,\xi) = T(t, \tilde{A}_0^*, \tilde{T}_0^*, \tilde{q}^*, \tilde{c}^*, \tilde{R}^*, \kappa, \xi) . \tag{3.3}$$

Consider the function

$$L(\kappa,\xi) = \frac{(t_2-t_1)}{n} \sum_{j=0}^{n} (\tilde{T}^* (\tau_j) - \tilde{T}(\tau_j,\kappa,\xi))^2 . \tag{3.4}$$

Let

$$\mu_n = \inf_{(\kappa,\xi)\varepsilon H^*} L(\kappa,\xi) . \tag{3.5}$$

Since L is continuous over the compact set H*, it follows that there exists a point $(\kappa^{(n)}, \xi^{(n)})$ in H* such that

$$\mu_n = L(\kappa^{(n)}, \xi^{(n)}) . \tag{3.6}$$

Lemma 2: *There exists a constant K which depends upon H and S such that*

$$\mu_n \le K^2 \varepsilon^2 . \tag{3.7}$$

Proof: See the proof [1] for the corresponding Lemma.

In Section 5 we will use Newton's Method to solve the nonlinear least-squares problem indicated in (3.5). Finally, we obtain the following result.

Theorem 2: There exist positive constants, K and ν, $0 < \nu < 1$, which depend only on H and S and such that

$$(3.8)\qquad \begin{aligned} |\kappa-\kappa^{(n)}| + |\xi-\xi^{(n)}| &\le K\ \{(K\left(\frac{t_2-t_1}{n} + \varepsilon^2\right) + 2\mu_n)^{\nu/2} + \varepsilon\} \\ &\le K\{(K\left(\frac{t_2-t_1}{n} + \varepsilon^2\right) + 2K^2\varepsilon^2)^{\nu/2} + \varepsilon\}\ . \end{aligned}$$

Proof: The proof is identical to the corresponding argument given in [1].

4. A Generalization to Systems

Consider the system of equations

$$(4.1)\qquad \begin{aligned} \dot{\vec{y}} &= \vec{F}(t, \vec{y}, \vec{k})\ , \\ \vec{y}(0) &= \vec{y}_0\ , \end{aligned}$$

where $\vec{y} = (y_1, y_2, \ldots, y_{m-1}, T)$ and $\vec{k} = (k_1, k_2, \ldots, k_p)$, and the function $\vec{F} = (F_1, F_2, \ldots, F_m)$ is analytic in all of its arguments. The problem for (4.1) that corresponds to the problem for (1.2) is that of determining a nonnegative $\vec{k}$ from the data $\vec{y}_0^*$ and $T^* = T^*(t)$ defined over the interval $t_1 \le t \le t_2$ $(t_1 \ge 0)$, where T^* is the mth component of the solution $\vec{y}$ of equation (4.1) with initial data $\vec{y}_0^*$.

From an integration of (4.1) it follows from the theory of ordinary differential equations that T is a continuous function of the initial data $\overline{y}_0$ and the vector $\vec{k}$:

$$(4.2)\qquad T = T(t, \vec{y}_0, \vec{k})$$

Definition: A solution of the problem for (4.1) is a nonnegative vector $\vec{k}$ such that for $t_1 \le t \le t_2$,

$$(4.3)\qquad T(t, \vec{y}_0^*, \vec{k}) = T^*(t)\ .$$

Now, we can use the concepts developed in Section 2 to obtain a condition for continuous dependence of the solution upon the data. We note that

$$\dot{T}(0, \vec{y}_0, \vec{k}) = F_m(0, \vec{y}_0, \vec{k}) . \tag{4.4}$$

The right hand side of (4.4) is a known function of $\vec{y}_0$ and $\vec{k}$, therefore, we can use the differential system to compute higher derivatives of T, for example,

$$\begin{aligned} \ddot{T} &= \frac{\partial F_m}{\partial t} + \sum_{j=1}^{m} \frac{\partial F_m}{\partial y_j} \dot{y}_j \\ &= \frac{\partial F_m}{\partial t} + \sum_{j=1}^{m} \frac{\partial F_m}{\partial y_j} F_j \end{aligned} \tag{4.5}$$

where in (4.5) we are using y_m to denote T. Also, we have that

$$\begin{aligned} \ddot{T}(0, \vec{y}_0, \vec{k}) &= \frac{\partial F_m}{\partial t}(0, \vec{y}_0, \vec{k}) \\ &\quad + \sum_{j=1}^{m} \frac{\partial F_m}{\partial y_j}(0, \vec{y}_0, \vec{k})\, F_j(0, \vec{y}_0, \vec{k}) \end{aligned} \tag{4.6}$$

and the right hand side of (4.6) is also a known function of $\vec{y}_0$ and $\vec{k}$. Now we denote $C_j(\vec{y}_0, \vec{k})$ by

$$C_j(\vec{y}_0, \vec{k}) = \frac{\partial^j y_m}{\partial t^j}(0, \vec{y}_0, \vec{k}) , \quad j = 1,2,\ldots,p \tag{4.7}$$

The C_j's are computed from the differential equations like (4.4) and (4.6) and are thus known functions. Following the pattern of Section 2, we consider

$$\begin{aligned} C_j(\vec{y}_0^{(2)}, \vec{k}^{(2)}) - C_j(\vec{y}_0^{(1)}, \vec{k}^{(1)}) &= \int_0^1 \frac{d}{d\theta} C_j(\vec{y}_0^{(1)}(1-\theta) + \vec{y}_0^{(2)}\theta , \\ &\qquad \vec{k}^{(1)}(1-\theta) + \vec{k}^{(2)}\theta) d\theta \quad , \end{aligned} \tag{4.8}$$

from whence it follows that

$$(4.9)\quad \sum_{i=1}^{p}\left(\int_0^1 \frac{\partial C_j}{\partial k_i}\,d\theta\right)(k_i^{(2)}-k_i^{(1)}) = D_j \ .$$

Thus, continuous dependence upon the data and unicity follow from the condition that

$$(4.10)\quad \det\left(\int_0^1 \frac{\partial C_j}{\partial k_i}\,(\vec{y}_0^{(1)}(1-\theta)+\vec{y}_0^{(2)}\theta,\ \vec{k}^{(1)}(1-\theta)+\vec{k}^{(2)}\theta)d\theta\right)\neq 0$$

for all $(\vec{y}_0^{(i)},\vec{k}^{(i)})$, $i=1,2$ in some suitable set, say

$$(4.11)\quad \begin{aligned} H = \{(\vec{y}_0,\vec{k})\,|\,\alpha_i \le y_{0i} \le \beta_i \quad &, \quad i = 1,2,\ldots,m \\ 0 < \gamma_j \le k_j \le \xi_j \quad &, \quad j = 1,2,\ldots,p\} \end{aligned}$$

of a priori bounds for the data $\vec{y}_0$ and the solution $\vec{k}$.

We note that for a particular nonlinear system, the structure of the system may permit simplifications of (4.10) as was the case with (2.2).

5. Numerical Experiments

In this section we implement the nonlinear least-squares method defined in Section 3. Specifically, we seek (k^*,E^*) such that

$$L(k^*,E^*) = \inf_{(k,E)} L(k,E)$$

$$(5.1)\quad \equiv \inf_{(k,E)} \sum_{j=0}^{n} (T^*(\tau_j) - T(\tau_j,A_0^*,T_0^*,q^*,c^*,R^*,k,E))^2\,\frac{(t_2-t_1)}{n},$$

where A_0^*, T_0^*, q^*, c^*, R^*, and $T^*(\tau)$ represent the data A_0 T_0, q, c, R and $T(\tau)$ with prescribed error. Our goal is to determine from calculation how these prescribed errors effect the estimates k^*, and E^*. In the implementation to follow, we assume that the thermodynamic variables (R=1.98, $q=10^5$, $c=10^{-4}$) are known without error--a reasonable assumption to make in practice since these constants can be accurately estimated by alternative methods which are not so prone to data measurement errors as is the case with the transient reaction system being observed here. The sequence of time measurement points

(5.2) $$\tau_j = j(t_2-t_1)/n \ , \quad j = 0,1,\ldots,n$$

was selected such that the time interval $[t_1,t_2]$ includes a significan portion of the transient period.

We are particularly interested in the effects on estimating k* and E* arising from three types of system measurement errors: (1) errors in initial conditions, (2) random errors in dynamic temperature measurement, and (3) systematic errors in temperature measurement arising say from "lag" or heat transfer effects associated with thermocouple temperature measurement.

In the following problem (k,E) was chosen to be $(10^{12},30000)$. These values of rate constant and activation energy are representative of those reported in the literature [5] for first-order gaseous reactions. The time interval over which discrete measurements were made is $0 \le t \le 5$. The initial conditions specified were

$$A_0 = 10^{-7}$$

$$T_0 = 500$$

(Units on k, E, A_0, T_0, t are respectively 1/sec, calories, mole/cm^3, degrees K and seconds.) For this problem, the system begins at an initial temperature of 500°K and rises to a steady state temperature of 600°K in about 5 seconds (see Fig. 1.).

To integrate system (1.3) through values τ_j, $j = 0,1,\ldots,n$ and thereby determine the "exact" temperature $T(\tau)$ corresponding to $A_0 = 10^{-7}$, $T_0 = 500$, $k = 10^{12}$, and $E = 30000$, we used the GEAR computer code [3], a Fortran subroutine for the integration of stiff ordinary matrix differential equations. High, controllable numerical accuracy is achieved at minimum computer expense by the dynamic variation of time steps and multistep integration methods used in the code.

In order to minimize L(k,E) in (5.1) with respect to the arguments k and E we seek roots of the nonlinear equation $\overline{P}(k,E) = 0$ where

(5.3) $$\overline{P}(k,E) = \begin{Bmatrix} \dfrac{\partial L}{\partial k} \\ \dfrac{\partial L}{\partial E} \end{Bmatrix} .$$

In order to solve this system, Newton's method was used. Use of Newton's method requires evaluation of

$$(5.4) \qquad T, \frac{\partial T}{\partial k}, \frac{\partial T}{\partial E}, \frac{\partial^2 T}{\partial k^2}, \frac{\partial^2 T}{\partial k \partial E}, \frac{\partial^2 T}{\partial E^2}$$

at the measurement points τ_j in (5.2). The transient behavior of the variables listed in (5.4) is determined by partial differentiation of equation (1.3) with respect to the parameters k and E together with the imposition of homogeneous initial conditions. Again, the GEAR code was used to carry out the integration of the 6 x 6 matrix differential equation governing the dynamics of the variables listed in (5.4).

We remark that the surface L(k,E) over the k-E plane has extremely elongated contours which cause the surface to look very much like a canyon with steep walls and a flat canyon floor. This in turn means that the observed temperature T(t) can be fit very well by parameters (k,E) which differ greatly from the exact values $k = 10^{12}$ and and $E = 30000$*. In the vicinity of the canyon floor, the derivatives $\frac{\partial^2 L}{\partial k^2}, \frac{\partial^2 L}{\partial k \partial E}$ are extremely small (of the order 10^{-15}). Consequently, the determinant of the Jacobian of $\overline{P}(k,E)$ is also small (of the order 10^{-15}) and the Newton calculations are very sensitive to rounding errors. This situation is avoided when we scale the problem by replacing the parameter k by $\overline{k} = \ln(k)$ and then estimating $\overline{k}$ and E by nonlinear least-squares. The estimate for k is then recovered from $\overline{k}$.

Remark:

The values of t_2, t_1 and n in (5.2) are important from a computational point of view. Formally, the only restrictions on these variables is that $0 \le t_1 \le t_2$ and $n \ge 2$ (since we are estimating two parameters k and E). Computational experiments on problems for which zero error was assumed in the data indicated that least-squares estimates k^* and E^* were insensitive to the values of t_1, t_2 and n with k^* and E^* agreeing with the true values ($k = 10^{12}$, $E = 30000$) to nine significant digits. However, to produce a convergent algorithm, small values of n ($n < 10$) or intervals $[t_1, t_2]$ that did not contain

* For example, the two solutions $T_1(t)$ and $T_2(t)$ to (1.3) determined by $k = 10^{12}$, $E = 30000$ and $k = 25 \times 10^{12}$, $E = 33153$; differ by less than .1% for all $t \ge 0$.

a significant portion of the transient period required initial guesses for Newton's method that were very close to the true least-squares solution. Larger values of n and intervals $[t_1,t_2]$ that embraced a substantial amount of the transient produced more robust Newton processes that were convergent from initial guesses which were generally within only 10% of the converged solution values. The values $n=25$ and the interval $[t_1,t_2]=[0,5]$ were used in the sequel.

5.1 Errors in Initial Data. To determine the effects of errors in initial data, T_0 and A_0, we nondimensionalize (1.3) by defining $T(t)/T_0$. With $T(t)$ now used to represent dimensionless temperature, we have from (1.3)

$$\dot{T} = k\exp(-E/RT_0T)\left(1 + \frac{qA_0}{cT_0} - T\right) \tag{5.5}$$

$$T(0) = 1\ .$$

Letting k^* and E^* be least-squares approximations to k and E in the presence of erroneous data T_0^*, A_0^*, then

$$\dot{T}(t,q,c,R,T_0^*,A_0^*,k^*,E^*) = k^*\exp(-E^*/RT_0^*T)\left(1 + \frac{qA_0^*}{cT_0^*} - T\right) \tag{5.6}$$

$$T(0) = 1\ .$$

It follows from (5.5) and (5.6) that if $\frac{qA_0^*}{cT_0^*} = \frac{qA_0}{cT_0}$, then $k^* = k$, $E^* = \left(\frac{T_0^*}{T_0}\right) E$, and $L(k^*,E^*) = 0$ (see row 5 of table 1). It also follows from (5.5) and (5.6) that if $\frac{qA_0^*}{cT_0^*} = \frac{qA_0^{**}}{cT_0^{**}}$, then $k^* = k^{**}$ and $E^* = \left(\frac{T_0^*}{T_0^{**}}\right) E^{**}$ (see, for example, row 2 and row 3 in Table 1). This implies that the least-squares estimate k^* is uniquely determined by the ratio $\frac{A_0^*}{T_0^*}$. Table 1 illustrates the numerical results of determining k^* and E^* in the presence of errors in A_0 and T_0. We remark that if it is the case that initial conditions cannot be regarded as exactly known, then they should assume the same role as the kinetic parameters k and E. They should be varied simultaneously with k and E in order to find the optimal set (k^*,E^*,T_0^*,A_0^*) which best fits the observed measurements (see [6]).

5.2 Random Errors in the Measurement of Temperature. In order to illustrate the effects of random errors on the estimates k* and E*, we first compute the exact temperature T(t) at the points τ_j. Then we perturb these calculations to define $T^*(\tau_j)$ in (5.1) by the addition of random perturbations:

$$T^*(\tau_j) = T(\tau_j)(1 + \alpha\, \xi_j) , \quad j = 0,1,\ldots,n$$

where ξ is drawn from a normal distribution with mean zero and a standard deviation of one. The value α in (5.5) controls the relative error in the temperature measurement:

In Table 2 we show numerical results of estimating k and E when different magnitudes, α, of relative error are assumed. Even for a relative error of 3% (as much as 15°K in 500°K), the estimate E* differs from the exact value by less than 2.5%.

5.3 Systematic Errors in Temperature Measurement. Usually, the temperature T(t) of a reacting system is monitored by a sensing and recording device (for example, a thermocouple and strip chart) and the reported value, say y(t), does not always if ever agree with T(t) exactly. Two simple causes of this discrepancy are system time delay (or "lag") and smoothing via convective heat transfer*. Under this assumption, the observed temperature y(t) can be related to the actual system temperature T(t) by the following equation:

$$\frac{dy(t)}{dt} = \frac{1}{\mu} (T(t-\tau) - y(t)) .$$

The coefficient $1/\mu$ in (5.6) is the effective convective heat transfer coefficient while τ represents a lag or time increment that is required for the temperature signal to reach the recording instrument**. For $\mu \to 0$, y(t) has the same graph as T(t) in Fig. 1 only its graph has been shifted to the right τ units (see Fig. 2).

For $\tau > 0$ and $\mu > 0$, y(t) will be less than T(t) for $t > 0$, hence the measuring system will consistently underestimate temperature. The consequence of least-squares fitting k and E based on the function y(t) is shown in Table 3. Included in Table 3 is a measure of the maximum relative error incurred by estimating T(t) by y(t):

---

* Time lag and heat transfer are not the only distortions that can occur, but they may be used to model others.

** In practice, when a thermocouple is used to measure temperature, μ falls in the range $.01 \le \mu \le 0.1$.

$$\|\cdot\| \equiv \max_{0 \le t} \left| \frac{T(t) - y(t)}{T(t)} \right| \times 100 \ .$$

Table 3 shows that underestimation of T(t) caused by lag or heat transfer and subsequent least-squares fitting to y(t) yields estimates of k and E that are too large. Once again, relative errors in estimates of the rate constant k are larger than corresponding relative errors in estimates of the activation energy, E for a given lag and heat transfer coefficient.

We conclude by remarking that in practice, it would generally be a mistake in estimating k and E to take the observed data y(t) and try to derive the actual temperature T(t) by

$$T(t) = y(t + \tau) + \mu \frac{dy}{dt} (t + \tau)$$

even if μ and τ were known with high confidence. This would involve differentiating experimental data which is not desirable. Rather, it is preferable to least-squares fit y(t) to the measured temperature data using parameters k, E, μ and τ and

$$\frac{dT}{dt} = k \exp(-E/RT) \left(T_0 + \frac{q}{c} A_0 - T\right)$$

$$\frac{dy}{dt} = \frac{1}{\mu} (T(t-\tau) - y(t)).$$

In this way, least-squares estimates of kinetic parameters k, E and measurement parameters μ, τ are developed simultaneously. After the best fit has been determined (k^*, E^*, μ^*, τ^*), the exact system temperature, T(t) can be recovered if desired.

6. Another Example.

6.1 Introduction. As in the previous sections the problem is to estimate the rate constant and energy of activation for a chemical reaction. The chemical reaction being considered here is:

$$A + B \xrightarrow{k} C + \text{heat} \tag{6.1}$$

which is described by the first order system of differential equations

$$\dot{A} = -k \exp(-E/RT)AB$$

$$\dot{B} = -k \exp(-E/RT)AB$$

$$\dot{C} = k \exp(-E/RT)AB$$

(6.2) $$\dot{T} = k \frac{q}{c} \exp(-E/RT)AB$$

$$A(0) = A_0$$

$$B(0) = B_0$$

$$C(0) = C_0$$

$$T(0) = T_0$$

where the initial concentrations A_0, B_0, and C_0 are nonnegative, the initial temperature T_0 is positive and q, c, and R represent respectively the heat of reaction, the system heat capacity, and the universal gas constant. The rate of reaction, k, and the energy of activation, E, are positive constants. The dot denotes differentiation with respect to the time variable t. Assuming the reaction is governed by the system of differential equations we are interested in estimating k and E given the initial data A_0^*, B_0^*, C_0^*, T_0^*, the positive constants q^*, c^*, and R^*, and a specified time interval $t_1 \le t \le t_2$, $t_1 \ge 0$. The transient temperature solution will be denoted by $T^*(t)$.

From the system of differential equations

$$\dot{T} + \frac{q}{c} \dot{A} = 0$$

and

$$\dot{T} + \frac{q}{c} \dot{B} = 0 .$$

This with some manipulation gives the differential equation

(6.3) $$\dot{T} = k \frac{c}{q} \exp(-E/RT)(T_0 + \frac{q}{c} A_0 - T)(T_0 + \frac{q}{c} B_0 - T)$$

$$T(0) = T_0$$

which T must satisfy.

It follows from the theory of solutions of differential equations that the solution

$$T = T(t,\ A_0,\ B_0,\ T_0,\ q,\ c,\ R,\ k,\ E)$$

depends continuously on the arguments t, A_0, B_0, T_0, q, c, R, k, and E.

Definition A solution of the Mathematical problem is a pair (k, ξ) such that for $t_1 \le t \le t_2$,

$$T(t,\ A_0^*,\ B_0^*,\ T_0^*,\ q^*,\ c^*,\ R^*,\ k, \xi\ ) \equiv T^*(t)\ .$$

Using the argument of section 2 we can show that the solution depends continuously upon the data. Let

$$\begin{aligned} H = \{ (A_0, B_0, T_0, q, c, R, k, \xi) :\ & A_{0*} \le A_0 \le A_0^{**}\ , \\ & B_{0*} \le B_0 \le B_0^{**},\ T_{0*} \le T_0 \le T_0^{**}\ ,\ q_* \le q \le q^{**}\ , \\ & c_* \le c \le c^{**}\ ,\ R_* \le R \le R^{**}\ ,\ K_* \le k \le K^{**}, \\ & \xi_* \le \xi \le \xi^{**} \}\ , \end{aligned}$$

and let S denote a compact set with nonempty interior which contains in its interior the closed interval $0 \le t \le t_2$ then the family of functions

$$\Omega = \{T(t, A_0, B_0, T_0, q, c, R, k, \xi) : (A_0, B_0, T_0, q, c, R, k, \xi) \in H,\ t \in S\}$$

is a family of analytic functions of the complex variable t.

We state now our result.

Theorem: If $(A_0^{(i)*},\ B_0^{(i)*},\ T_0^{(i)*}, q^{(i)*}, c^{(i)*}, R^{(i)*}, k^{(i)}, \xi^{(i)})$ $i = 1,2$ belong to H then there exist constants K and ω, $0 \le \omega \le 1$ which depend on H and S such that

$$\begin{aligned} |k^{(2)} - k^{(1)}| + |\xi^{(2)} - \xi^{(1)}| \le K\ \{ & \|T_2^* - T_1^*\|^{\omega} + |q^{(2)*} - q^{(1)*}| \\ & + |A^{(2)*} - A_0^{(1)}| + |B_0^{(2)*} - B_0^{(1)*}| + |T_0^{(2)*} - T_0^{(1)*}| \\ & + |c^{(2)*} - c^{(1)*}| + |R^{(2)*} - R^{(1)*}| \}\ . \end{aligned}$$

From this, we have uniqueness.

Proof: See the analysis of Section 2. The details are left to the reader.

6.2 Numerical Experiments.

Nonlinear least squares was used to make approximations to the solution of this problem as in Section 5. Attention was restricted to experiments where the initial approximation to the solution was the solution. (i.e. $K = 10^{16}$, $E = 30{,}000$). This allowed Newton's method to be used to obtain the approximation since the initial approximation was the "exact" solution.

The values of R, c, and q were set at 1.98, 10^{-4}, and 10^{5} respectively. The initial concentrations A_0 and B_0 were set at 10^{-7} and the initial temperature T_0 was set at 600° .

The range of temperatures was from 600° to approximately 773° in a time span of 1/4 sec. This was not the entire transient. Within 0.004 sec. after this span the temperature more than doubled to approximately 1700°.

When all values were exact the routine iterated once and gave back $E = 29{,}999$ and $K = 0.99999 \times 10^{16}$ which is less than 0.01% relative error in the approximation.

Error was introduced into the data using the form

$$T(t_j) = T(t_j)(1 + \alpha\gamma_j) \qquad j = 1,2,\dots, n$$

where α controls the relative error and γ_j's are drawn from a random number generator in the program. Table 4 shows the results for different values of α .

TABLE 1

Least-squares estimates k*, E* in the presence of initial condition errors. Exact parameter values are $A_0 = 10^{-7}$, $T_0 = 500$, $k = 10^{12}$, $E = 30{,}000$.

A_0^*	T_0^*	T*	k*	E
0.99×10^{-7}	500	T	1.36×10^{12}	30,307
1.01×10^{-7}	500	T	$.72 \times 10^{12}$	29,685
1.00×10^{-7}	495	T	$.72 \times 10^{12}$	29,385
1.00×10^{-7}	505	T	1.36×10^{12}	30,610
0.99×10^{-7}	495	T	1.00×10^{12}	29,700

TABLE 2

Least-squares estimates k*, E* in the presence of random error, $T^* = T(1 + \alpha\xi)$. Exact values are $k = 10^{12}$, $E = 30{,}000$.

A_0^*	T_0^*	α	T*	k*	E*
1×10^{-7}	500	.005	T + $\frac{1}{2}$ % random	$.84 \times 10^{12}$	29,823
1×10^{-7}	500	.01	T + 1 % random	$.73 \times 10^{12}$	29,681
1×10^{-7}	500	.02	T + 2 % random	$.60 \times 10^{12}$	29,470
1×10^{-7}	500	.03	T + 3 % random	$.52 \times 10^{12}$	29,320

TABLE 3

Least-squares estimates of k* and E* in the presence of systematic error (τ = lag, $\frac{1}{\mu}$ = heat transfer). Exact values are k = 10^{12}, E = 30,000.

A_0^*	T_0^*	T*	$\frac{1}{\mu}$	τ	$\|\cdot\|$	k*	E*
$1\text{x}10^{-7}$	500	$y(\mu,\tau)$	10	0	1.8%	$1.33\text{x}10^{12}$	30,330
$1\text{x}10^{-7}$	500	$y(\mu,\tau)$	100	0	0.3%	$1.09\text{x}10^{12}$	30,091
$1\text{x}10^{-7}$	500	$y(\mu,\tau)$	∞	.1	1.9%	$1.84\text{x}10^{12}$	30,663
$1\text{x}10^{-7}$	500	$y(\mu,\tau)$	∞	.2	3.6%	$3.25\text{x}10^{12}$	31,279
$1\text{x}10^{-7}$	500	$y(\mu,\tau)$	80	.01	0.5%	$1.17\text{x}10^{12}$	30,168

TABLE 4

Note: The initial concentrations were taken to be exact values ($A_0 = B_0 = 10^{-7}$) and T_0 was taken to be exact ($T_0 = 600°$)

α	T*	k*	E*
0.01	T + 1% Random	1.000×10^{16}	29,999
0.05	T + 5% Random	1.000×10^{16}	29,980
0.10	T + 10% Random	1.000×10^{16}	29,970
0.50	T + 50% Random	1.000×10^{16}	27,860

In these experiments for the system (6.2) the error tolerance in the GEAR program was extremely small (10^{-14}) which would account for a little more accuracy. Note also the effect of $k^* = 10^{16}$ versus 10^{12} in Section 5. One might conclude that large rate constants and activation energies can be found accurately in the face of large random error in the temperature measurements. A note of caution should be injected here since the solution curves to such systems of ordinary differential equations are essentially the same for large rate constants and activation energies. Quite probably, Newton's method saw no intrinsic reason to move away from the exact k* and E* due to the flatness of the valley floor for the nonlinear least squares method.

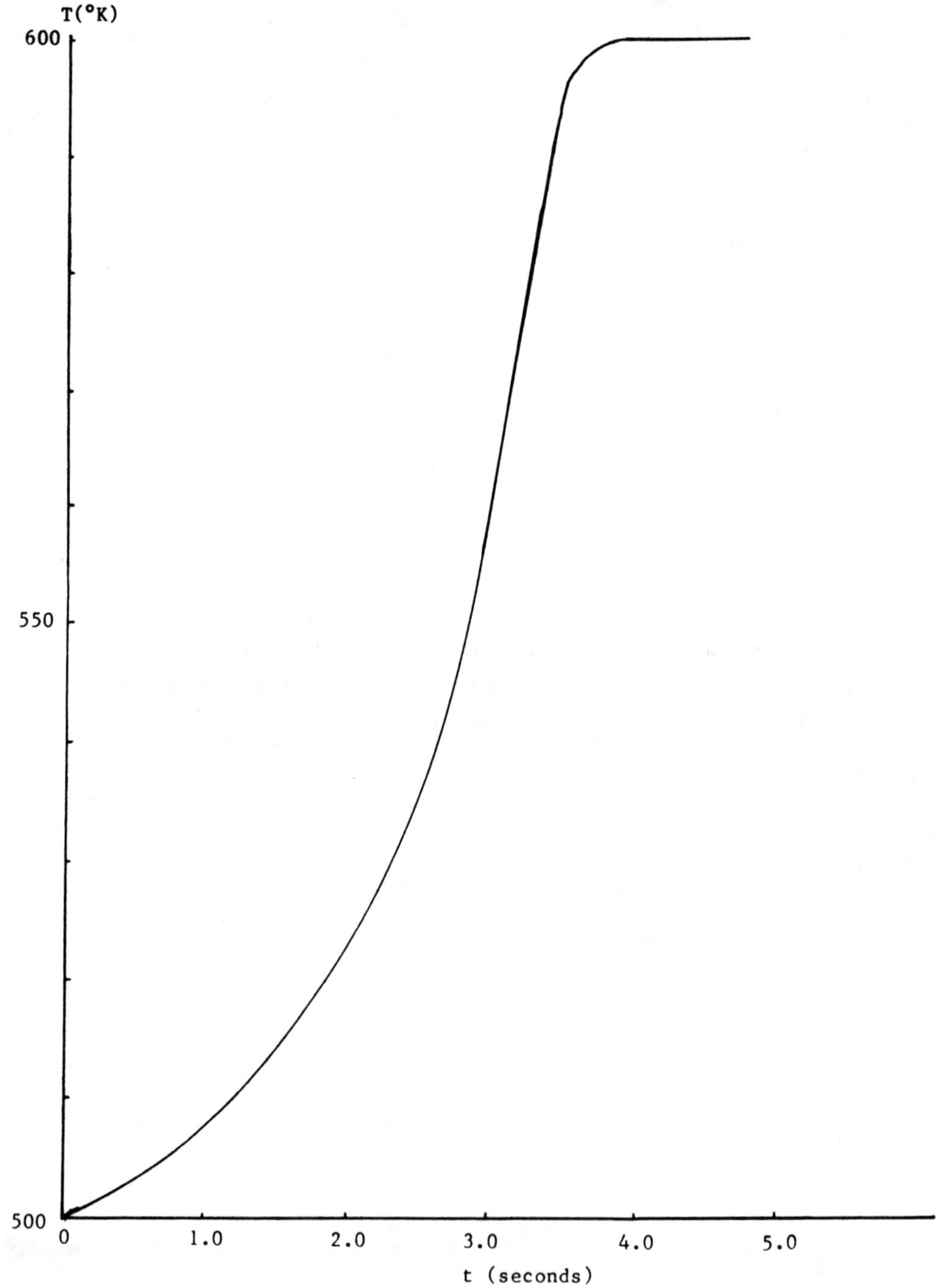

Figure 1. Temperature versus time. $A_0 = 10^{-7}$, $T_0 = 500$, $k = 10^{12}$, $E = 30000$

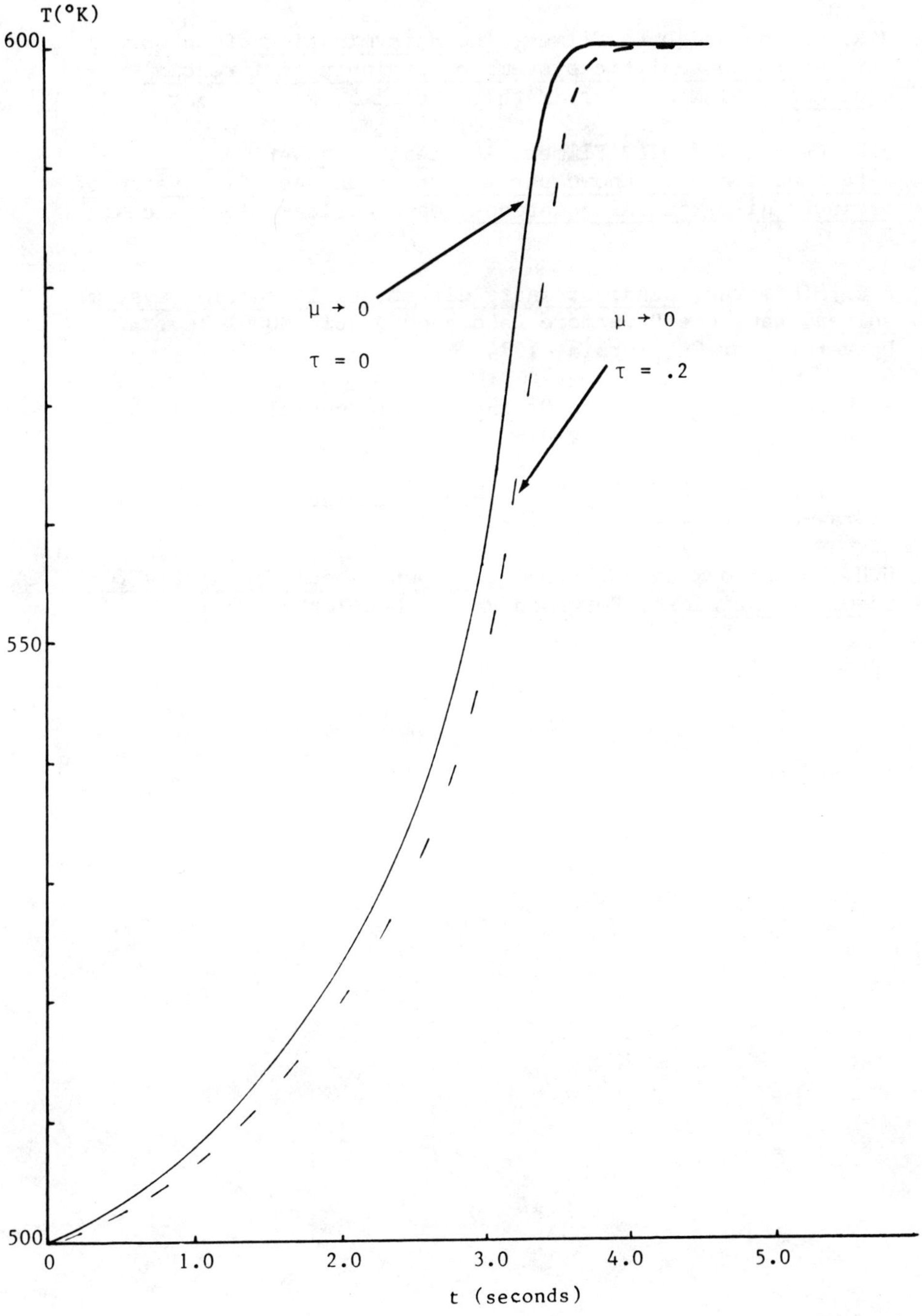

Figure 2. The effect of lag on temperature

REFERENCES

[1] J.R. Cannon and D.L. Filmer, The determination of unknown parameters in analytic systems of ordinary differential equations, SIAM J. Appl. Math., Vol. 15, No. 4, July 1967.

[2] J.R. Cannon and D.L. Filmer, A numerical experiment on the determination of unknown parameters in an analytic system of ordinary differential equations, Mathematical Biosciences, Vol. 3, 1968.

[3] A.C. Hindmarsh, GEAR: ordinary differential equation system solver, Lawrence Livermore Laboratory, UCID-30001 Rev. 3, University of California, 1974.

[4] W. Hurewicz, Lectures on Ordinary Differential Equations, John Wiley & Sons, Inc., New York, 1958.

[5] R.H. Perry and C.H. Chilton, Chemical Engineers Handbook, McGraw-Hill Book Company, 1973.

[6] H.H. Rosenbrock and C. Storey, Computational Techniques for Chemical Engineers, Pergamon Press, New York, 1966.

CONTINUATION TECHNIQUES IN THE STUDY OF CHEMICAL REACTION SCHEMES

EUSEBIUS DOEDEL*

Abstract. We describe various continuation techniques that can be useful in the computer-aided analysis of dynamical systems. Of particular interest are stationary and periodic phenomena in autonomous systems of ordinary differential equations. Software that has been developed for this purpose is described. Detailed results are presented for the application of the techniques to an exothermic reaction scheme involving two consecutive reactions.

1. Introduction. Continuation techniques can be very useful in the numerical analysis of mathematical models that contain one or more parameters. See for example [1,7,18,24,28,29,30,36]. Numerical methods are treated in [8,21,22,27,33,35,41]. The objective is often to determine all possible solutions that the equations can have as one or more parameters are varied. Here we describe some of these techniques and we apply them to a chemical reaction scheme.

In Section 2 we consider continuation techniques for autonomous systems of ordinary differential equations: First we recall the continuation of stationary solutions and periodic solutions in one parameter. Extensions to delay differential equations and to the computation of invariant curves of discrete dynamical systems are mentioned. Periodically forced systems and infinite period orbits (homoclinic or heteroclinic) are also discussed. Then we indicate how steady state limit points, Hopf bifurcation points, period solution limit points, period doubling bifurcations, and bifurcations to invariant tori can be continued in two parameters. Extensions to parabolic systems are discussed also.

The implementation of the basic schemes into general software (AUTO) is described in Section 3. We detail the capabilities of a current version of the program (July 1983), and of an extended version which is not yet in final form at the time of this writing. In Section 4, which can be read independently, we apply some of these continuation methods to the computer-aided classification of chemical reactor behaviour.

* Computer Science Department, Concordia University, Montréal, Canada.
Supported in part by NSERC Canada (#4274) and FCAC Québec (#EQ1438).

The results illustrate how part of the analysis of certain dynamical systems can be done on the computer using continuation methods. The exothermic reaction scheme considered involves consecutive reactions $A \rightarrow B \rightarrow C$, and has been studied previously in [7,16,19,25]. Reference [7] includes the computation of entire periodic solution branches with both stable and unstable solutions. Earlier in [16] Hopf bifurcations in this reaction scheme were analyzed. In [25] certain period doubling sequences and chaotic regimes are discribed in great detail. Our numerical analysis in Section 4 brings to light a number of interesting phenomena, including multiple stable periodic behaviour and isolas of periodic solutions. The two-parameter continuations illustrate various interactions of Hopf bifurcation points, stationary limit points, homoclinic orbits, and periodic solution limit points. A general classification of the behaviour resulting from a certain class of such interactions can be found in [12]. (See also [26,31].)

2. Continuation Techniques for Autonomous ODE's. We consider the autonomous differential equation

$$u'(t) = f(u(t), \lambda), \quad u, f(\cdot, \cdot) \in R^n . \tag{2.1}$$

Steady State Continuation. The computation of steady state solution branches of (2.1) is an algebraic problem consisting of the bifurcation analysis of the equation.

$$f(u, \lambda) = 0 . \tag{2.2}$$

This can be done numerically by applying the general continuation and branch switching techniques of H.B. Keller [27]. The continuation method inflates the original problem (2.2) by adding a normalization. The inflated system is

$$\begin{aligned} &f(u, \lambda) = 0 , \\ &\theta_1^2 (u - u_0)^* \dot{u}_0 + \theta_2^2 (\lambda - \lambda_0) \dot{\lambda}_0 - \Delta s = 0 , \end{aligned} \tag{2.3}$$

where Δs (a "pseudo-arclength") replaces λ as the continuation parameter. Above $(\dot{u}_0, \dot{\lambda}_0)$ denotes the direction of the branch at the preceding solution point (u_0, λ_0), " * " denotes transpose, and θ_1 and θ_2 are weights. One advantage of using the inflated system (2.3) is the capability of computing past limit points on the solution branch.

Continuation of Periodic Solutions. Along branches of periodic solutions of autonomous systems, both the solution $u(t)$ and its period T changes as λ varies. To fix the period we scale the time variable t and transform (2.1) into

(2.4) $$u'(t) = T f(u(t) , \lambda), t \in [0,1] .$$

Now the unknown period T appears explicitly, and we seek solutions satisfying

(2.5) $$u(0) = u(1) .$$

Thus the initial value problem has been replaced by a boundary value problem with nonseparated boundary conditions, which allows the computation of unstable orbits and computation near singular points [8,21,36,41]. Suppose we have computed a solution triple $(u_0(\cdot), T_0 , \lambda_0)$. The objective is to set up the equations for computing a next solution $(u(\cdot), T , \lambda)$ at "distance" Δs from the given solution. Thus the pseudo arclength normalization is now defined by

(2.6) $$\theta_1^2 \int_0^1 (u(t) - u_0(t))^* \dot{u}_0(t) \, dt + \theta_2^2 (\lambda - \lambda_0)\dot{\lambda}_0 + \theta_3^2 (T - T_0) \dot{T}_0 - \Delta s = 0 .$$

We still have to add an "anchor" equation (phase condition) which eliminates free translation of u(t) in time. Although there are many possible choices for such an equation, the following is very convenient for problems with rapidly varying solution components:

(2.7) $$\int_0^1 u(t)^* u_0'(t) \, dt = 0 .$$

For a discussion of the effect of choice of the anchor equation see [8]. Equations (2.4), (2.5) and (2.7) define a bifurcation problem of the form

$$F(w , \lambda) = 0 ,$$

where $w \equiv (u , T)$ and to which the normalization (2.6) is still to be added. Again the methods of [27] can be used to switch branches at ordinary bifurcation points and also at period doubling bifurcations. Such bifurcations can be located accurately by our software since the Floquet multipliers are extracted as part of the solution procedure. We do not give details on this here, although a brief discussion of the discretization is included in Section 3.

<u>Delay Differential Equations and Discrete Dynamical Systems</u>. The above method for computing stable and unstable orbits can be extended to delay differential equations. For one implementation see [9]. The bifurcation analysis of delay differential equations is also considered in [15,38,39]. Another interesting extension concerns the continuation of invariant curves in discrete dynamical systems. These can arise from a Hopf bifurcation in a diffeomorphism. A method that produces the rotation number as a byproduct is contained in a recent

thesis [6]. We have written software for delay and difference equations, but it is not yet as fully developed as that for the numerical analysis of (2.1).

Periodically Forced Systems. These can often be rewritten as autonomous systems. This is convenient if software for bifurcation analysis is only available for autonomous systems (e.g. AUTO). The idea is to add a nonlinear oscillator to the system, that has the desired periodic forcing as one of its solution components. An example of such an oscillator is

$$x' = x + \theta y - x (x^2 + y^2) ,$$

$$y' = -\theta x + y - y (x^2 + y^2) ,$$

which has the asymptotically stable solution $x = \sin(\theta t)$, $y = \cos(\theta t)$ as output. As an example of a driven system consider the forced Duffing equation:

$$u' = v ,$$

$$v' = -u - \lambda (u^3 + \alpha y) ,$$

where αy represents the forcing term. Thus the original forced system (with $y = \cos(\theta t)$) has been replaced by the enlarged but autonomous system. The important parameters are the frequency θ and the coupling strength α. A more direct approach to the numerical bifurcation analysis of a forced Duffing equation can be found in [4]. For a wealth of results on forced oscillations the reader is referred to [17]. Very complicated bifurcation behaviour in a system of linearly coupled oscillators is analyzed in [1].

Orbits of Infinite Period. In (2.4)— (2.7) we can fix the period T, e.g., $T = T^0$, and free another parameter, say μ. Thus we have

$$u'(t) = T^0 f(u(t) , \lambda , \mu) ,$$

$$u(0) = u(1) .$$

$$\theta_1^2 \int_0^1 (u(t) - u_0(t))^* \dot{u}_0(t)\, dt + \theta_2^2 (\lambda - \lambda_0) \dot{\lambda}_0$$

$$+ \theta_3^2 (\mu - \mu_0) \dot{\mu}_0 - \Delta s = 0 .$$

$$\int_0^1 u(t)^* u_0'(t)\, dt = 0 .$$

The above system can then be used for computing curves $(u(\cdot), \lambda , \mu)$ of fixed period orbits. If the period is very high then the curves approximately represent curves of homoclinic (or heteroclinic) orbits. Using our very accurate and reliable adaptive discretization it is in fact possible to continue orbits with extremely high fixed period in two parameters. For an application see [1].

Continuation of Steady State Limit Points. Let the differential equation be

$$u'(t) = f(u, \lambda, \mu), \quad u, f(., ., .) \in R^n, \lambda, \mu \in R,$$

where λ and μ are two free parameters. The steady state condition is

$$f(u, \lambda, \mu) = 0, \tag{2.8}$$

and a singular point with one dimensional null space satisfies

$$\begin{aligned} f_u(u, \lambda, \mu)v &= 0, \quad v \in R^n \\ v^* v - 1 &= 0. \end{aligned} \tag{2.9}$$

Equations (2.8) and (2.9) define an algebraic continuation (or bifurcation) problem of 2n+1 equations in 2n+2 variables, viz., u, v, λ and μ. Except for degenerate cases, one will have curves of solutions in 2n+2-dimensional space. This algebraic problem can be analysed numerically by the same techniques used for the continuation of steady states. For a detailed discussion of efficient techniques for the continuation of limit points see [13,22,35] and references therein. The continuation of limit points naturally leads to the detection of cusp singularities when these are present. The cusp can then be continued in three parameters, etc. [23]. In fact we have carried out such computations for the reaction scheme of Section 4, but we do not report the results here.

Continuation of Hopf Bifurcation Points. As in equation (2.4) let the differential equation (with scaled time variable) be given by

$$u'(t) = T f(u, \lambda, \mu), \tag{2.10}$$

where λ and μ are the two free parameters. Again the steady state condition is

$$f(u, \lambda, \mu) = 0. \tag{2.11}$$

A necessary condition for Hopf bifurcation from a branch of steady states is that at the point of bifurcation the linearized problem

$$v'(t) = T f_u(u, \lambda, \mu) v(t),$$

have, for some value of T, a nonconstant solution v(t) satisfying

$$v(0) = v(1) = 0.$$

If $f_u(u, \lambda, \mu)$ has a simple pair of purely imaginary eigenvalues with imaginary part $1/T$ then the solution of the linearized problem has the form

$$v(t) = \sin(2\pi t)\,\xi + \cos(2\pi t)\,\eta .$$

Substituting this into the linearized equation and separating the sine and cosine terms results in the two equations

$$(2.12)\qquad \begin{aligned} \frac{T}{2\pi} f_u\, \xi + \eta &= 0 , \\ -\xi + \frac{T}{2\pi} f_u\, \eta &= 0 , \end{aligned}$$

where $f_u \equiv f_u(u, \lambda, \mu)$. We normalize the null vector (ξ, η) of (2.12) by requiring

$$(2.13)\qquad \xi^* \xi + \eta^* \eta - 1 = 0 .$$

However the nullspace is two-dimensional, due to the freedom of translating $v(t)$ in time. This requires another constraint. Let $v_0(t) \equiv \sin(2\pi t)\,\xi_0 + \cos(2\pi t)\,\eta_0$ correspond to the previous solution point on the branch of Hopf bifurcations. To anchor $v(t)$ we require that

$$d(\sigma) \equiv \int_0^1 (v(t+\sigma) - v_0(t))^* (v(t+\sigma) - v_0(t))\, dt$$

be minimized over σ. This results in the equation

$$\int_0^1 (v(t) - v_0(t))^*\, v_0'(t)\, dt = 0 .$$

Substitution of the expressions for $v(t)$ and $v_0(t)$ into the above equation, followed by integration, gives

$$(2.14)\qquad \eta_0^T \xi - \xi_0^T \eta = 0 .$$

The equations (2.11)— (2.14) define an algebraic bifurcation problem of the form

$$(2.15)\qquad F(w, \lambda) = 0 ,$$

where now $w \equiv (u, \xi, \eta, T, \mu) \in R^{3n+2}$. Here n is the dimension of the differential equation. The notation suggests that λ has been chosen as the "free" parameter, but this choice is of course unimportant. Rather, $F(w, \lambda) = 0$, represents a nonlinear system of $3n+2$ equations in $3n+3$ unknowns. Thus generally one will have curves of solutions to this equation, with possible limit points and bifurcation points. Very similar schemes for the continuation of Hopf bifurcation points were also derived in [14,21]. Curves of Hopf bifurcation points may have limit points which can then be continued in three parameters, etc. We have done such computations for the $A \rightarrow B \rightarrow C$ reaction scheme in Section 4, but the results will not be discussed here.

Continuation of Periodic Limit Points. As illustrated in Section 4, more complete insight into the oscillatory behaviour of a system can be obtained by the two parameter continuation of limit points on branches of periodic solutions. From a general point of view the formulation of this continuation problem is identical to that given above for continuing steady state limit points. If we again let

$$F(w, \lambda, \mu) = 0, \quad w \equiv (u, T), \tag{2.16}$$

represent the equations (2.4), (2.5) and (2.7) for the continuation of periodic solutions, then a limit point on a branch of periodic solutions necessarily satisfies

$$F_w(w, \lambda, \mu)\, n = 0, \quad n \equiv (v, \sigma), \tag{2.17}$$

where the null "vector" **n** may be normalized by requiring

$$\|\mathbf{n}\|^2 - 1 = 0, \tag{2.18}$$

in an appropriate norm. Thus the equations (2.16)–(2.18) are the analogue of the equations (2.8),(2.9) for the continuation of limit points on stationary solution branches. Numerically, of course, the equations (2.16)–(2.18) are somewhat more complex to deal with. Writing out these equations in full detail, using the definition of the aggregated operator F, we find

$$u'(t) - T f(u, \lambda, \mu) = 0$$

$$u(0) - u(1) = 0$$

$$\int_0^1 u(t)^* u_0'(t)\, dt = 0$$

$$v'(t) - T f_u(u, \lambda, \mu) v - \sigma f(u, \lambda, \mu) = 0$$

$$v(0) - v(1) = 0$$

$$\int_0^1 v(t)^* u_0'(t)\, dt = 0$$

$$\int_0^1 v(t)^* v(t)\, dt + \sigma^2 - 1 = 0 .$$

Again one can write these equations in general operator form as

$$G(x, \lambda) = 0, \tag{2.19}$$

where now $x \equiv (u, v, T, \sigma, \mu)$. Thus equation (2.19) represents a system of ordinary differential equations with nonseparated boundary conditions and with additional integral constraints and a corresponding number of scalar unknowns. Still to be added is the normalization that allows the computation of solution branches to (2.19) to proceed around limit points. We omit writing this normalization in full detail here. A different approach to the continuation of limit points on branches of periodic solutions is considered in [36], p. 47. Much efficiency can be gained by making use of the fact that the singularity of the system is really a property of the Poincaré map [34]. In the current implementation we have not made use of this yet.

Continuation of Period-Doubling Bifurcations and Tori. Replacing the boundary condition $v(0) - v(1) = 0$ by $v(0) + v(1) = 0$ in the equations represented by (2.19) gives the scheme for continuing period doubling bifurcations in two parameters. One can also derive quite easily an extended system for the two parameter continuation of bifurcations to invariant tori. In this case one will have a system of 3n first order differential equations, subject to 3n non-separated boundary conditions and 3 integral constraints (plus pseudo arclength continuation). There will be four scalar unknowns also. Since an invariant torus can be characterized as an invariant curve of the Poincaré return map, the latter can be continued also using the continuation of invariant curves for discrete maps as underlying algorithm. However the actual implementation, using reliable numerical techniques, is fairly complex and work on this is still in progress.

Continuation Techniques for Parabolic Systems. Consider nonlinear diffusive systems with time variable t and one space variable x of the form

$$u_t = u_{xx} + f(x, u, u_x, \lambda), \qquad x \in [0,1], \quad t \geq 0, \tag{2.20}$$

where $u(x, t), f(\cdot, \cdot, \cdot, \cdot) \in R^n$, $\lambda \in R$. This nonlinear parabolic system is subject to 2 n boundary conditions of the form

$$b_0(u(0), u_x(0), \lambda) = 0, \qquad b_1(u(1), u_x(1), \lambda) = 0. \tag{2.20a}$$

For notational simplicity we have assumed that the diffusion term is linear with diffusion constants equal to one. Numerical bifurcation analyses of chemical reactors modelled by equations of the above type can be found in, e.g., [18,20].

One can discretize (2.20) in space, after which the system (2.20), (2.20a) reduces to an autonomous system of ordinary differential equations. This system then can be analyzed by the techniques of Section 2. Two major problems arise in this approach.

Firstly, systems with intermediate and high spatial structure will require an accurate approximation scheme in space. Invariably this will result in ordinary differential systems of high dimension for which the numerical bifurcation analysis will require large amounts of computer time. Secondly, even with a relatively accurate (fixed) discretization in space, the ordinary differential system readily exhibits parasitic solution branches. Thus, while it may be possible to analyze the ordinary differential equations without difficulty, the solution structure found may be quite different from that of the full partial differential equation. Our experience indicates, that this approach ("the method of lines" in time) results in meaningful bifurcation diagrams primarily in those situations, where the model without diffusion (ODE's, CSTR) could have been used to begin with. Nevertheless, some reliable results using the above approach have been obtained for certain problems, including the Brusselator equations (where tori were found) and an enzyme model with diffusion [30].

The difficulties mentioned above arise especially in the analysis of time periodic phenomena. For this reason we generally do not use the numerical techniques for the continuation of periodic solutions in the numerical analysis of partial differential equations. Rather, transient computations can be used to obtain insight into the time-dependent behaviour.

Parabolic Systems: Steady State Continuation. For the steady state bifurcation analysis of (2.20),(2.20a) we can use the numerical continuation and bifurcation techniques. The stationary equations are

$$u'' + f(x, u, u', \lambda) = 0, \tag{2.21}$$

subject to

$$b_0(u(0), u'(0), \lambda) = 0, \qquad b_1(u(1), u'(1), \lambda) = 0. \tag{2.21a}$$

Here and below $'$ denotes differentiation with respect to x. The equations (2.21) represent a nonlinear two-point boundary value system with free parameter λ. These equations can be analyzed using the techniques of [27] together with accurate and adaptive discretization techniques. Aspects related to discretization will be taken up in Section 3.

Parabolic Systems: Continuation of Steady State Limit Points. We can interpret equations (2.21),(2.21a) as an operator equation. A limit point on a branch of stationary solutions necessarily is a singular point of the operator equation. Thus, completely analogously to equations (2.8),(2.9), and also (2.16),(2.17) and (2.18), we obtain the following equations for the continuation of steady state limit points in two parameters:

$$u'' + f(x, u, u', \lambda, \mu) = 0,$$

$$b_0(\, u(0)\,, u'(0)\,,\, \lambda\,,\, \mu\,) \;=\; 0\,,$$

$$b_1(\, u(1)\,, u'(1)\,,\, \lambda\,,\, \mu\,) \;=\; 0\,,$$

$$v'' + f_u(\, x\,, u\,, u'\,,\, \lambda\,,\, \mu\,)\, v + f_{u'}\,(\, x\,, u\,, u'\,,\, \lambda\,,\, \mu\,)\, v' \;=\; 0\,,$$

$$b_{01}(\, u(0),\, u'(0),\, \lambda\,,\, \mu\,)\, v\,(0) + b_{02}(\, u(0),\, u'(0),\, \lambda\,,\, \mu\,)\, v'(0) = 0\,,$$

$$b_{11}(\, u(1),\, u'(1),\, \lambda\,,\, \mu\,)\, v\,(1) + b_{12}(\, u(1),\, u'(1),\, \lambda\,,\, \mu\,)\, v'(1) = 0\,,$$

$$\int_0^1 v(t)^*\, v(t)\, dt - 1 = 0\,.$$

Above b_{01}, b_{02}, b_{11} and b_{12} represent the partial derivatives of b_0 and b_1 with respect to their first and second arguments. The above expanded system represents 2 n second order ordinary differential equations subject to 4 n boundary conditions and an integral constraint. In addition to the vector valued functions u and v, there are also two scalar unknowns, namely λ and μ. (The pseudo-arclength equation still needs to be added.) Thus the continuation of stationary limit points in the two free parameters λ and μ requires the numerical continuation of solutions to a system of dimension essentially twice that of the original stationary system (2.21) , (2.21a). Alternate methods that lead to a much smaller expanded system are under current investigation (e.g. [34]). The expanded system may have limit points itself (which corresponds to cusps or to the birth of isolas) or even bifurcation points. The latter are more likely to occur if the system has certain symmetries.

Parabolic Systems: Continuation of Hopf Bifurcation Points. The continuation of Hopf bifurcation points in two parameters for systems of parabolic partial differential equations is the immediate generalization of the same procedure for ordinary differential equations. We can interpret (2.1) , (2.11)– (2.14) as general operator equations, and as operator we take the differential operator in (2.21) with boundary conditions (2.21a). Writing out the resulting equations in full we obtain:

$$u'' + f(\, x\,, u\,, u'\,,\, \lambda\,,\, \mu\,) \;=\; 0\,,$$

$$b_0(\, u(0)\,, u'(0)\,,\, \lambda\,,\, \mu\,) \;=\; 0\,,$$

$$b_1(\, u(1)\,, u'(1)\,,\, \lambda\,,\, \mu\,) \;=\; 0\,,$$

$$\frac{T}{2\pi}\,\{\, \xi'' + f_u\,(\, x\,, u\,, u'\,,\, \lambda\,,\, \mu\,)\, \xi$$

$$+ f_{u'}(\, x\,, u\,, u'\,,\, \lambda\,,\, \mu\,)\, \xi'\,\} + \eta \;=\; 0\,,$$

$$b_{01}(u(0),u'(0),\lambda,\mu)\xi(0)+b_{02}(u(0),u'(0),\lambda,\mu)\xi'(0)=0\ ,$$

$$b_{11}(u(1),u'(1),\lambda,\mu)\xi(1)+b_{12}(u(1),u'(1),\lambda,\mu)\xi'(1)=0\ ,$$

$$-\xi+\frac{T}{2\pi}\{\eta''+f_u(x,u,u',\lambda,\mu)\eta$$
$$+f_{u'}(x,u,u',\lambda,\mu)\eta'\}=0\ ,$$

$$b_{01}(u(0),u'(0),\lambda,\mu)\eta(0)+b_{02}(u(0),u'(0),\lambda,\mu)\eta'(0)=0\ ,$$

$$b_{11}(u(1),u'(1),\lambda,\mu)\eta(1)+b_{12}(u(1),u'(1),\lambda,\mu)\eta'(1)=0\ ,$$

$$\int_0^1\{\xi(t)^*\xi(t)+\eta(t)^*\eta(t)\}\,dt-1=0\ ,$$

$$\int_0^1\{\eta_0^*(t)\xi(t)-\xi_0^*(t)\eta(t)\}\,dt=0\ .$$

The above expanded system represent 3 n second order ordinary differential equations, subject to 6 n boundary conditions and two integral constraints. In addition to the unknown vector valued functions u, ξ, and η, there are another three scalar unknowns, namely the period T and the parameters λ and μ. This system will in general have curves of solutions ($u(x\ ;\ s)$, $\xi(x\ ;\ s)$, $\eta(x\ ;\ s)$, $T(s)$, $\lambda(s)$, $\mu(s)$), where s is some parametrization. Again there may be limit points (with respect to λ, say) and bifurcations.

3. Software. We have developed a subroutine package (AUTO) that can do the continuation and bifurcation computations described in Section 2. A first package for the one-parameter stationary and periodic bifurcation analysis of the system (2.1) was already completed in August 1979. It was presented at the June 1980 SIAM meeting in Alexandria Va., and at the August 1980 Workshop on Boundary Value Problems in Vancouver, B.C. A first written account appeared in [8]. Results of the application of the original program to the exothermic A → B → C reaction is reported in [7] and results of applying AUTO to a system describing linearly coupled nonlinear oscillators in contained in [1]. Applications to enzyme models are included in [10,11,30]. The program has been made available (on an informal basis) to interested academic users. Further distribution requires permission of the author. The latest version that has been made available (July 1983) can do the stationary and periodic bifurcation analysis of the system (2.1). More specifically, the program can trace out branches of steady states and accurately locate limit points, steady state bifurcation points and Hopf bifurcation points. It can switch onto bifurcating steady state and periodic solution branches. Floquet multipliers are computed along periodic solution branches and secondary periodic bifurcations can be located. Switching branches is possible at regular secondary periodic

bifurcation points (transcritical and pitch fork bifurcations) and at period doubling bifurcations. Reliable computation of periodic orbits is possible due to the high order collocation methods with adaptive mesh selection that are used in the program. The two parameter continuation schemes in the July 1983 version are confined to the computation of curves of steady state limit points and Hopf bifurcation points.

An extended version, which at present is not yet sufficiently documented to be made available, consists of the following principal parts:

(1) A collection of subroutines for the bifurcation analysis of nonlinear algebraic systems. This includes branch tracing, detection of limit points and bifurcation points, and branch switching.

(2) Subroutines for the bifurcation analysis of very general first order systems of ordinary differential equations subject to nonlinear (or linear) boundary conditions and integral constraints. The number of side conditions need not match the order of the differential equation if there is a corresponding number of additional scalar unknowns.

(3) Subroutines for the continuation and bifurcation analysis of periodic solutions of (2.1).

The continuation schemes described in Section 2 can be realized in this structure. In fact most of these have been built into the extended version. (Delay differential equations, invariant curves of discrete dynamical systems, and the continuation of tori are not yet included. At present these are treated separately, On the other hand some continuation schemes for the optimal control of multistate systems are already an integral part of the extended version.)

The software has been designed to minimize programming effort when dealing with a given equation. Thus for the stationary and periodic bifurcation analysis of an autonomous system of the form (2.1) one need only supply the vector-valued function f and its Jacobian f_u and the derivative f_λ. The expanded system required for the multi-parameter continuation of stationary limit points, Hopf bifurcation points, periodic limit points, etc. is generated automatically when needed. Also the starting procedure requires little user intervention. For example, if a Hopf bifurcation has been found in the course of a one-parameter bifurcation analysis, then information concerning the bifurcation is written into a file. The bifurcation point is also assigned a label. To continue the Hopf bifurcation in two parameters one need only restart the program, indicating that a two-parameter continuation is to be done and indicating the label of the Hopf bifurcation point to be retrieved.

The extended version has a preprocessor in order to minimize the size of the load module. The preprocessor reads the user-supplied routines to determine what is wanted (e.g. a bifurcation analysis of an algebraic system, the restarting of the computation of previously computed branch of orbits, the two parameter continuation of a Hopf point, etc.). The workspace (array space), which depends on the application and on the accuracy of the discretization, is also computed. The preprocessor then generates the main program, inserting the appropriate calls as well as the required dimension of the workspace. This main program is then compiled and loaded along with the user-supplied routines and the required subroutines from the AUTO-library.

Discretization of differential equations is required for (2) and (3). In AUTO the method of orthogonal collocation with piecewise polynomials is used [7,8]. The number of Gauss collocation points in each subinterval can be selected. There is a general collocation program for ordinary boundary value problems [2,3], but it has not been developed for the numerical analysis of bifurcation problems. For this reason it was necessary to write a separate collection of subroutines to deal with the discretization. (For another general package for boundary value problems in ordinary differential equations see [32].) A few ideas were borrowed from [2,3], specifically the criterion used for distributing the mesh [37]. A good strategy for adapting the mesh is obviously useful in keeping down the number of degrees of freedom in the discretization. Use of an adaptive mesh is also to some extent a safeguard against parasitic solution branches. (See for example [5] for a discussion of the effect of discretization on bifurcation diagrams.) This can be explained as follows: If an insufficiently accurate discretization is used, then its bifurcation diagram may differ considerably from that of the continuous problem. It will generally also differ from the diagram of other discretizations, e.g. the diagram for a different mesh distribution. If an adaptive mesh selection is used when continuing a branch of solutions, then for each adaption one actually switches from one bifurcation diagram to another. If the two are not "close", then convergence difficulties will be encountered, since an initial approximation to the next solution is always obtained by extrapolation from previously computed solutions on the branch. Even if convergence is obtained, the branch may have a "jagged" appearance, indicating an inadequate discretization. (The latter situation is frequently observed near homoclinic orbits of (2.1).) Either of these two situations will signal that the discretization is not accurate enough and that the number of mesh intervals or the number of collocation points per interval has to be increased. (For proving the existence of orbits using computer-based techniques see [40]).

4. <u>The $A \to B \to C$ Reaction</u>. We consider consecutive, first order reactions in an ideal stirred tank, continuously fed by component A and cooled by a medium with constant temperature. The mass balances for A and B and the energy balance, in dimensionless form, are written

below [7,16,19]:

$$
\begin{aligned}
& y' = -y + D\,(1 - y)\,\exp(\theta)\,, \\
(4.1)\quad & z' = -z + D\,(1 - y)\,\exp(\theta) - D\,\sigma\,z\,\exp(\theta) \\
& \theta' = -\theta - \beta\,\theta + D\,B\,(1 - y)\,\exp(\theta) + D\,B\,\alpha\,\sigma\,z\,\exp(\theta).
\end{aligned}
$$

Here $1 - y$ is the concentration of A, z the concentration of B, θ the temperature, D the Damkohler number, σ the selectivity ratio, β is the heat transfer coefficient, B is the adiabatic temperature rise and α is the ratio of heats of reaction.
There are five parameters in (4.1). An exhaustive classification is difficult to obtain and will certainly require a very lengthy description. To simplify matters we fix three of the parameters: $\alpha = 1.0$, $\sigma = 0.04$ and $B = 8.0$. Thus there are two free parameters left, viz. D and β. Other computations that we have done seem to indicate that the results below are typical.

The results of the numerical analysis using the continuation techniques of Section 2 as implemented in AUTO are summarized in Figures 1-14. Figures 1-3 are two parameter diagrams showing the various curves of stationary limit points, periodic limit points and Hopf bifurcation points. Figures 4-14 are one-parameter bifurcation diagrams with the Dankohler number as bifurcation parameter.

In the one-parameter bifurcation diagrams the vertical axis is the norm of the solution. For stationary solutions this norm is the Euclidean norm of the steady state solution vector. For periodic solutions the norm is defined as

$$
\left(\frac{1}{T}\right)^{\frac{1}{2}} \left\{ \int_0^T u(t)^* \, u(t)\, dt \right\}^{\frac{1}{2}},
$$

where T is the period of the oscillation and where u(t) describes the orbit in R^3. With these definitions a Hopf bifurcation point occupies the same point in the bifurcation diagram, whether considered as a point on a stationary branch or as a point on a periodic branch. Also if a branch of periodic solutions approaches a homoclinic orbit (infinite period) as its terminal point, then in the bifurcation diagram this terminal point coincides with the stationary saddle point that lies on the homoclinic orbit.

Figure 1 shows the curve of stationary limit points in the D - β plane. This curve was computed by the continuation method for stationary limit points indicated in Section 2. To obtain a starting point on this curve, an initial one parameter bifurcation diagram having stationary limit points was sought. Such a particular stationary limit point is located very accurately and information

concerning the point is written in a file. Setting a simple switch will cause AUTO to retrieve this information and to start continuing the curve of limit points in two parameters.
In the right margin of Figure 1 (and similarly in Figures 2 and 3) we have indicated the Figure numbers that contain the one-parameter bifurcation diagrams corresponding to the indicated cross section. Thus, for example, the cross sections in Figure 1 refer to the bifurcation diagrams in Figures 4,5,6,7,8 and 10. Indeed, there are two stationary limit points in Figures 4 and 5, four limit points in Figures 6 and 7, one in Figure 8, and none in Figure 10. Three cusps can be seen in Figure 1. Two of these are relatively close to each other, indicating the proximity of a swallowtail singularity. (We have also continued the cusps in three parameters, by freeing σ as third parameter, but these results will be omitted here).

The following simple conclusions can be drawn from Figure 1: For values of D and β corresponding to points outside the region bordered by the curve of limit points, the differential equation admits only a unique stationary solution. Points inside this region, but not inside the swallowtail, yield three stationary solutions. Finally, a point inside the swallowtail structure corresponds to parameter values for which there are five stationary solutions.
A variety of jump phenomena are suggested by Figure 1 when continuously varying both D and β. However some care must be taken in making assumptions about the stability of the stationary solutions involved. The presence of Hopf bifurcations may alter the dynamic behaviour considerably. In fact this is what happens in the equations under investigation here.

Figure 2 shows the curve of Hopf bifurcation points in the D - β plane. Again this curve has been obtained by the continuation scheme for Hopf bifurcation points described in Section 2. A starting point on the curve was found from a one-parameter analysis. As was the case for limit points, the user involvement in the details of the continuation and starting procedures is minimal. (We have also continued the limit points (with respect to β) on the curve of Hopf bifurcation points in Figure 2. The third parameter was σ. Results are omitted here). We see that for $\beta = 1.00$ there is only one Hopf bifurcation (see Figure 4), for $\beta = 1.25$ there are two (Figure 6), while for $\beta = 1.50$ there are four (Figure 10).
Note that the curve of Hopf bifurcation points in Figure 2 contains two endpoints, one near $D = 0.156$, $\beta = 1.24$ and the other near $D = 5.5\ 10^{-4}$, $\beta = 0.18$. These endpoints are singular points in the sense that the period T tends to infinity when approaching the end of the curve. The period referred to here is the period at the Hopf bifurcation point. This period exists only in a limiting sense and is proportional to the reciprocal of the imaginary part of the purely imaginary eigenvalues of the corresponding stationary solution. Hence the stationary solution associated with the endpoint must have at least a double eigenvalue at zero and is therefore singular. In fact, the

endpoints of the curves of Hopf bifurcations in Figure 2 lie on the curves of limit points in Figure 1. This could be observed if the two figures were superimposed. We have avoided this and other superimpositions here in order to retain interpretable graphs.
We stress the distinction between the infinite period bifurcation found above (Hopf bifurcation with infinite period at the Hopf bifurcation point) and homoclinic orbits (nontrivial orbit, containing a saddle point, with infinite period). The latter is sometimes also referred to as an infinite period bifurcation. To add to the confusion, the two types are often found close to each other. More carefully stated, if one takes a one-parameter cross-section of Figure 2 near, but not at, an endpoint of the curve of Hopf bifurcation points, then the corresponding one-parameter bifurcation diagram will contain "short" branches of periodic solutions. Such a short branch originates from a Hopf bifurcation point that lies close to a stationary limit point and terminates in a homoclinic orbit. As an example consider Figure 6 (blow-up), where a very short branch can be seen near the first stationary limit point. Figure 4 provides another example. A recent analysis of this situation can be found in [26].

The curve of Hopf bifurcation points, as shown in Figure 2, does not provide us with complete information concerning the periodic behaviour of the two parameter model. This behaviour is also dependent on the presence of secondary periodic bifurcation points and periodic solution limit points. In the parameter range considered here, the periodic solution limit points are of primary interest. As an example, for the same number of Hopf bifurcation and stationary limit points, two quite different bifurcation diagrams can be seen in Figures 10 and 14.

To start the discussion of the global periodic behaviour, consider Figure 9. In agreement with Figure 2, there are four Hopf bifurcations. There are also two stationary limit points in the bifurcation diagram of Figure 9, which agrees with Figure 1. The local enlargement given in Figure 9 reveals the presence of three periodic solution limit points. This is already interesting in its own right, since it gives rise to a hysteresis phenomenon with periodic solutions. (See also [7]). Using the continuation scheme for limit points on branches of periodic solutions that was introduced in Section 2 we are able to continue these limit points. The second parameter is again β, and the results of this continuation process are displayed in Figure 3. A local enlargement is shown in Figure 3a.

Next consider the behaviour that occurs for β between the values 1.54 and 1.62. This region of $D - \beta$ plane is shown enlarged in Figure 3a. The one-parameter bifurcation diagrams that show cross-sections for various values of β in this region are given in Figures 11, 12, 13 and 14.
We start with Figure 11. There are four Hopf bifurcation points. A branch of periodic solutions connects the first Hopf point to the fourth. Another branch connects the second and third Hopf bifurcation point.

In Figure 14, however, the connections have changed: the first point connects to the second, and the third connects to the fourth. By what mechanism did the interchange of connections occur? This question is answered by considering the diagrams for intermediate values for β which are shown in Figures 12 and 13. Effectively the interchange takes place between Figures 11 and 12. At some value of β the growing inner branch of periodic solutions meets the outer branch and the connections change in the usual manner of perturbed bifurcation. The exact point where this occurs corresponds to the limit point (with respect to β) on the branch of periodic solution limit points near $D = 0.252$, $\beta = 1.563$ in Figure 3a.
Another very similar interchange takes place between Figures 12 and 13 and results in an isola of periodic solutions. Again the exact location where this second interchange of connections takes place can be found in Figure 3a. It corresponds to the β– limit point near $D = 0.223$, $\beta = 1.571$ on the lower curve of periodic solution limit points. In this case increasing β results in the disappearance of two periodic solution limit points.

The upper portion of the isola of periodic solutions in Figure 13 consists of asymptotically stable solutions. We note that these oscillations are not likely to be observed when varying the Damkohler number and when β is fixed at $\beta = 1.58$. This likelihood is even less for $\beta = 1.60$ where the isola has shrunk in size (See Figure 14). If both parameters D and β can be varied, then the stable oscillations on the upper part of the isola can of course, be reached via a continuous path. For increasing β the isola further decreases in size and finally disappears. This occurs at the β– limit point near $D = 0.24$, $\beta = 1.617$ on the upper branch of periodic solution limit points in Figure 3a.

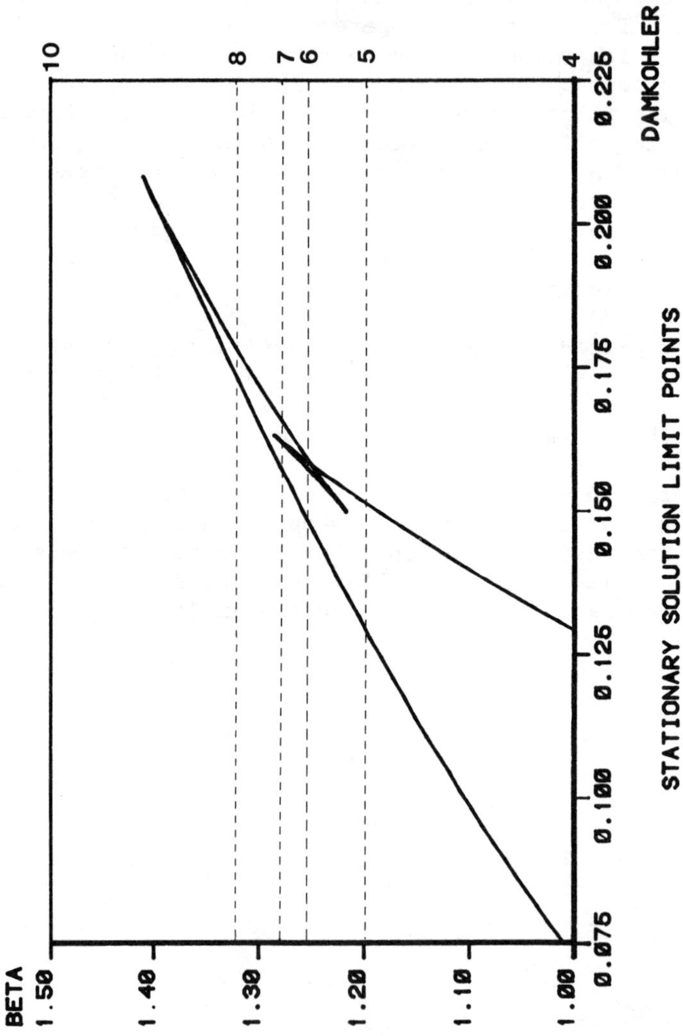
BETA
1.50
1.40
1.30
1.20
1.10
1.00
0.075
0.100
0.125
0.150
0.175
0.200
0.225
DAMKOHLER
10
8
7
6
5
4
STATIONARY SOLUTION LIMIT POINTS

FIGURE 1

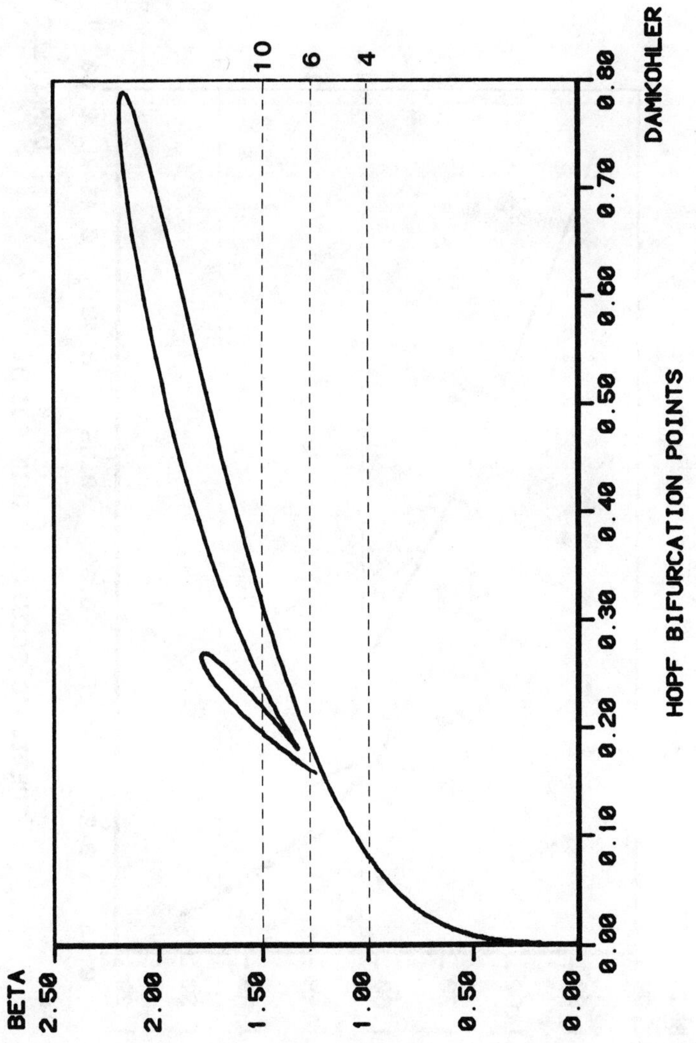

BETA
2.50
2.00
1.50
1.00
0.50
0.00
0.00
0.10
0.20
0.30
0.40
0.50
0.60
0.70
0.80
DAMKOHLER
10
6
4
HOPF BIFURCATION POINTS

FIGURE 2

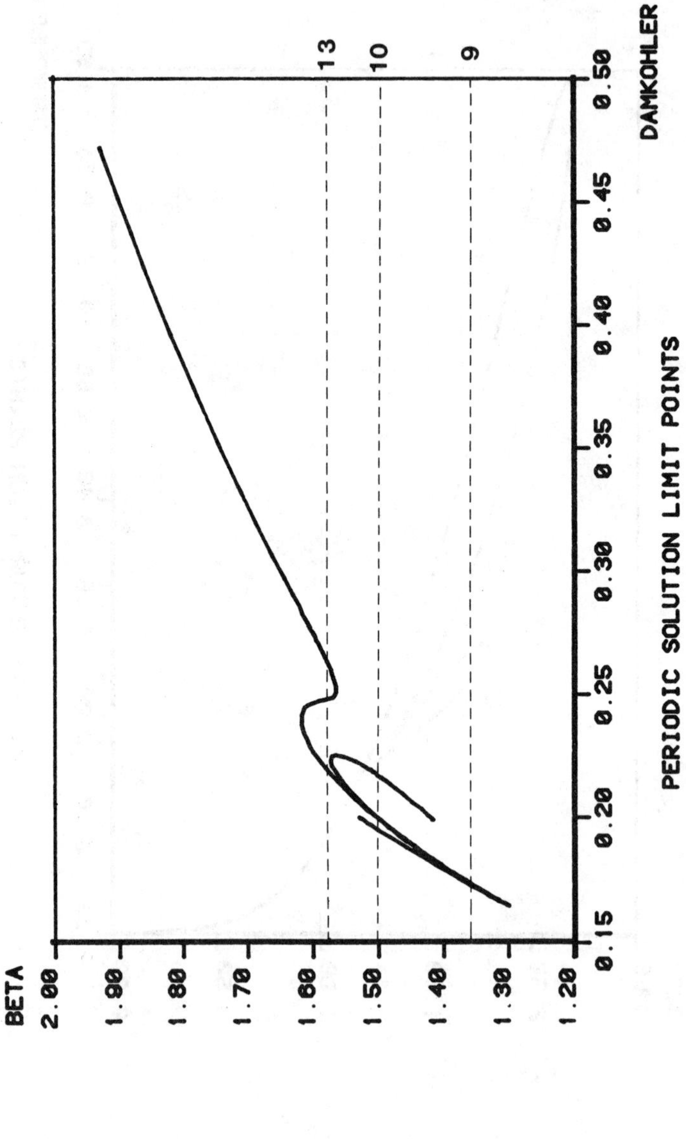
BETA
2.00
1.90
1.80
1.70
1.60
1.50
1.40
1.30
1.20
0.15
0.20
0.25
0.30
0.35
0.40
0.45
0.50
DAMKOHLER
13
10
9
PERIODIC SOLUTION LIMIT POINTS

FIGURE 3

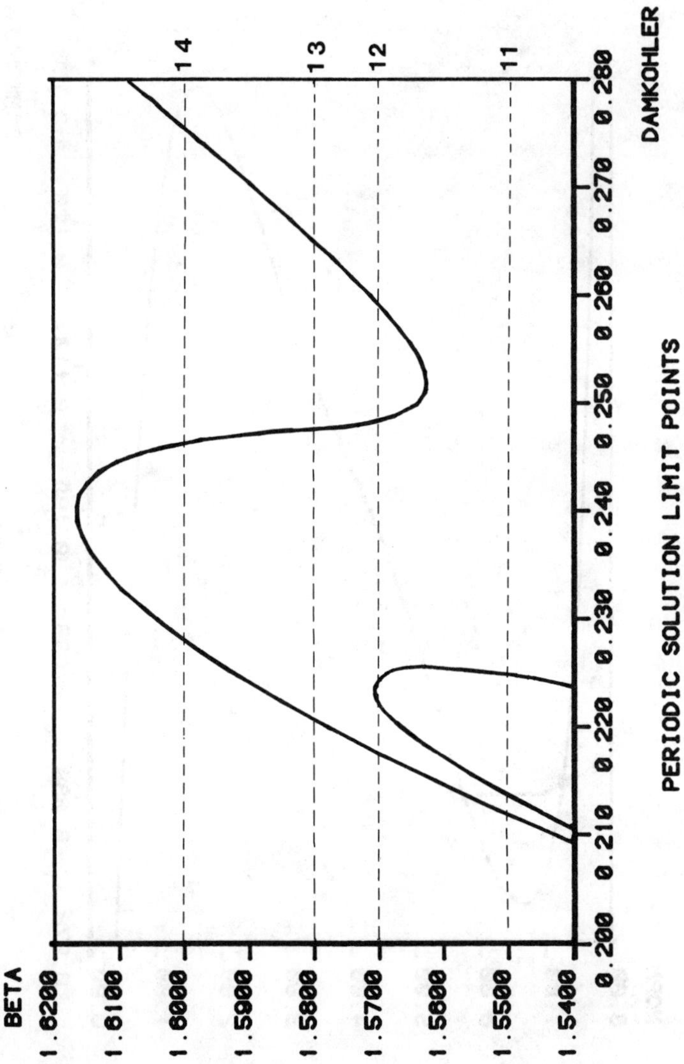
BETA
1.6200
1.6100
1.6000
1.5900
1.5800
1.5700
1.5600
1.5500
1.5400
0.200
0.210
0.220
0.230
0.240
0.250
0.260
0.270
0.280
DAMKOHLER
14
13
12
11
PERIODIC SOLUTION LIMIT POINTS

FIGURE 3A

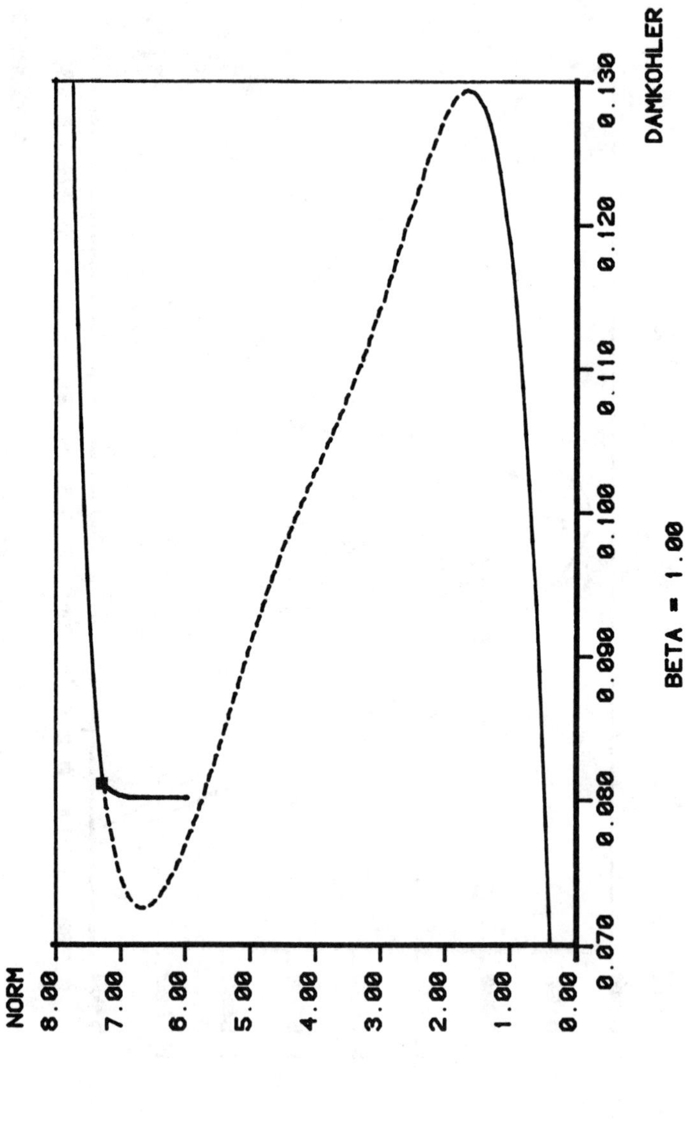
NORM
8.00
7.00
6.00
5.00
4.00
3.00
2.00
1.00
0.00
0.070
0.080
0.090
0.100
0.110
0.120
0.130
DAMKOHLER
BETA = 1.00

FIGURE 4

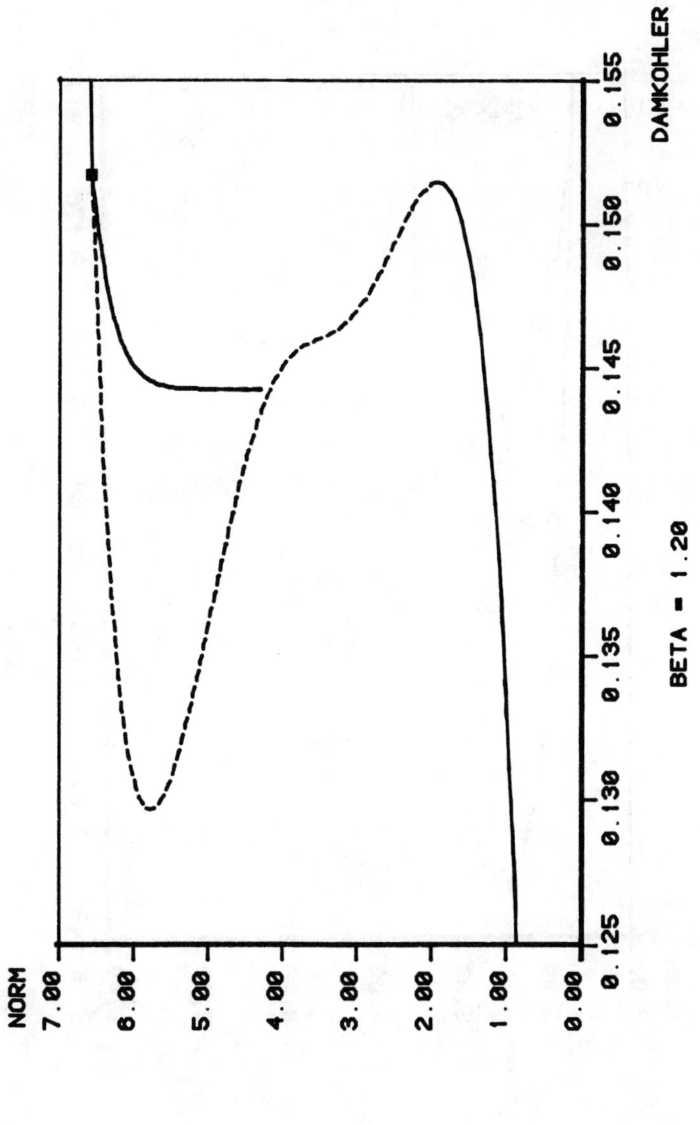
NORM
7.00
6.00
5.00
4.00
3.00
2.00
1.00
0.00
0.125
0.130
0.135
0.140
0.145
0.150
0.155
DAMKOHLER
BETA = 1.20

FIGURE 5

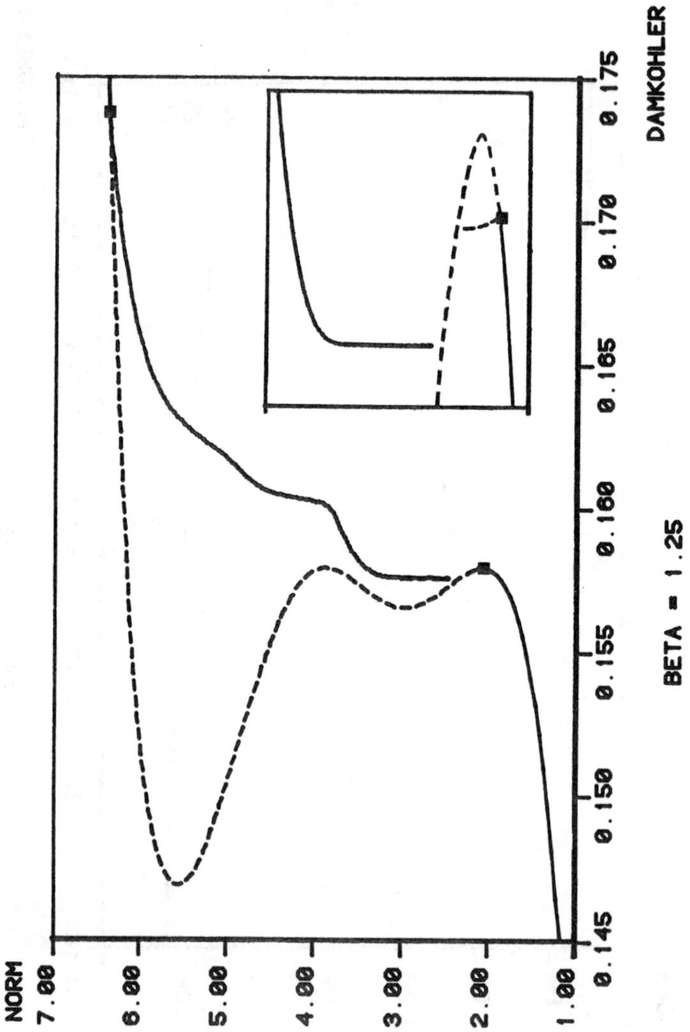
NORM
7.00
6.00
5.00
4.00
3.00
2.00
1.00
0.145
0.150
0.155
0.160
0.165
0.170
0.175
DAMKOHLER
BETA = 1.25

FIGURE 6

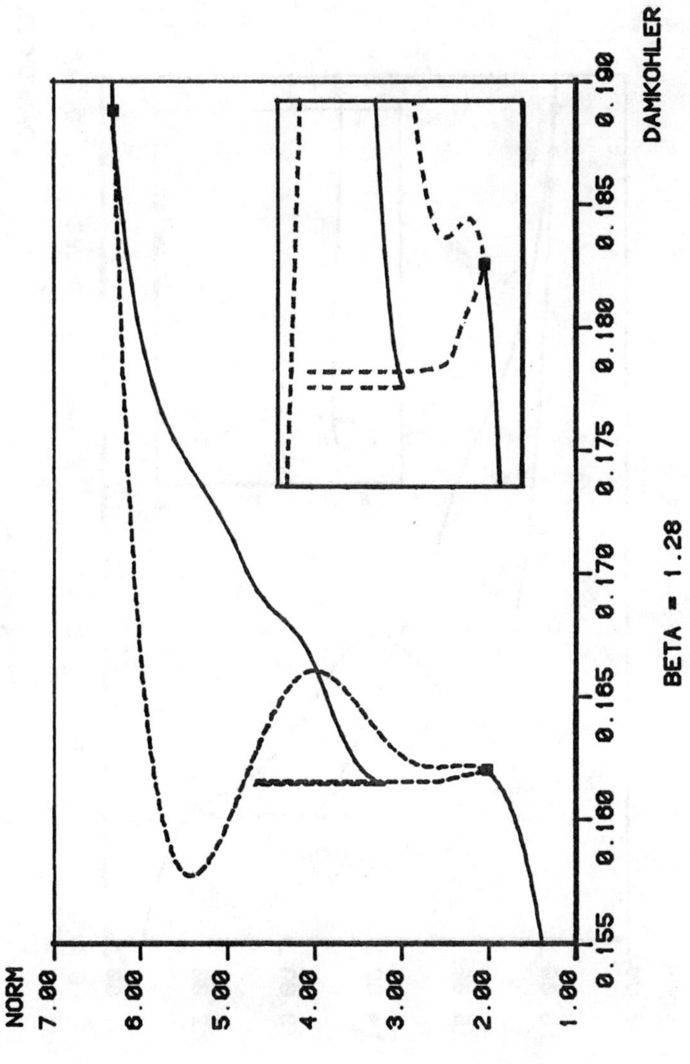
NORM
7.00
6.00
5.00
4.00
3.00
2.00
1.00
0.155
0.160
0.165
0.170
0.175
0.180
0.185
0.190
DAMKOHLER
BETA = 1.28

FIGURE 7

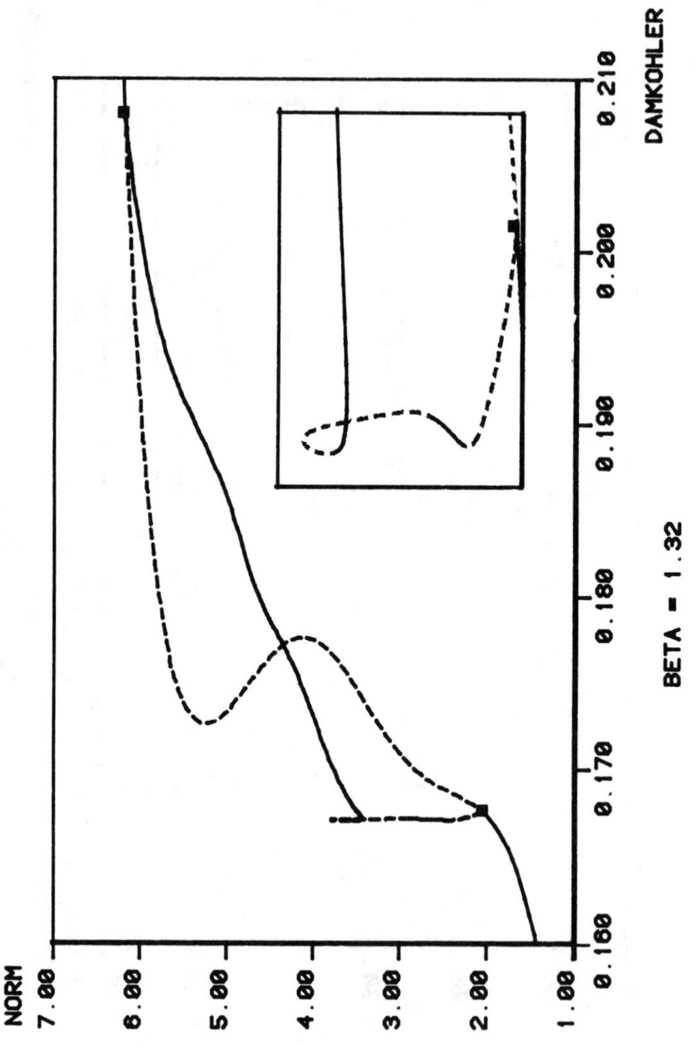
NORM
7.00
6.00
5.00
4.00
3.00
2.00
1.00
0.160
0.170
0.180
0.190
0.200
0.210
DAMKOHLER
BETA = 1.32

FIGURE 8

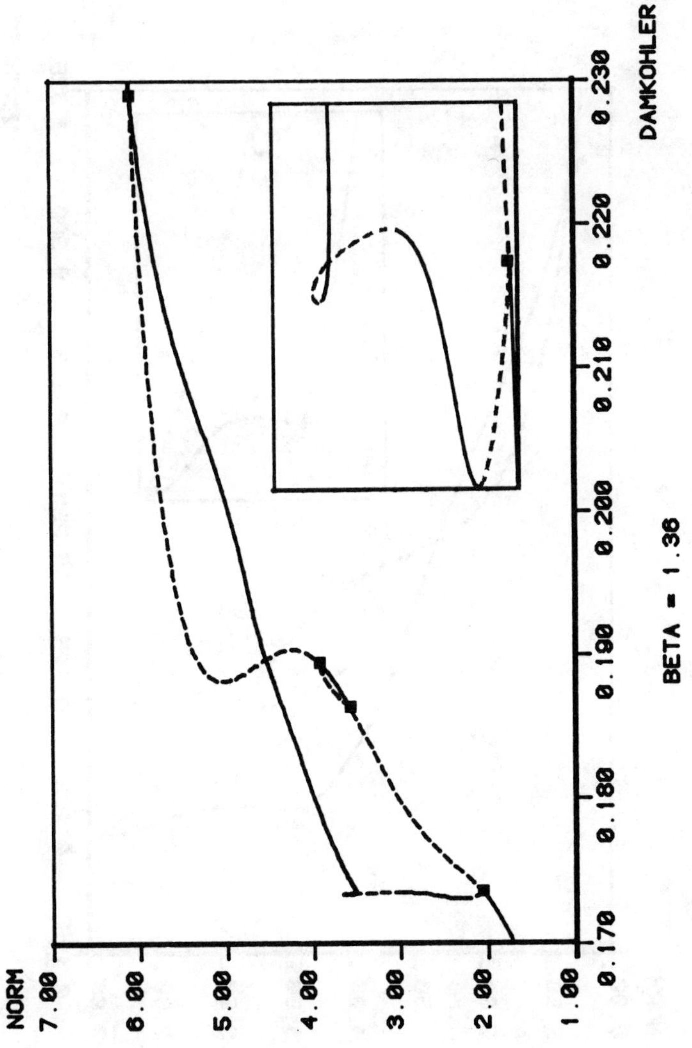
NORM
7.00
6.00
5.00
4.00
3.00
2.00
1.00
0.170
0.180
0.190
0.200
0.210
0.220
0.230
DAMKOHLER
BETA = 1.36

FIGURE 9

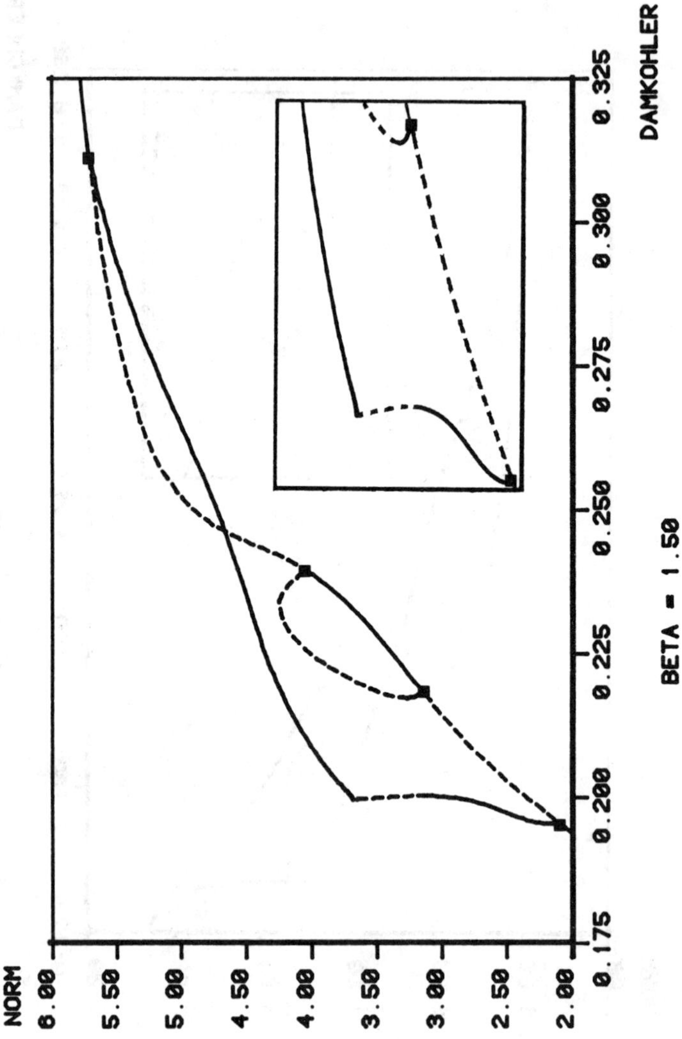
NORM
6.00
5.50
5.00
4.50
4.00
3.50
3.00
2.50
2.00
0.175
0.200
0.225
0.250
0.275
0.300
0.325
DAMKOHLER
BETA = 1.50

FIGURE 10

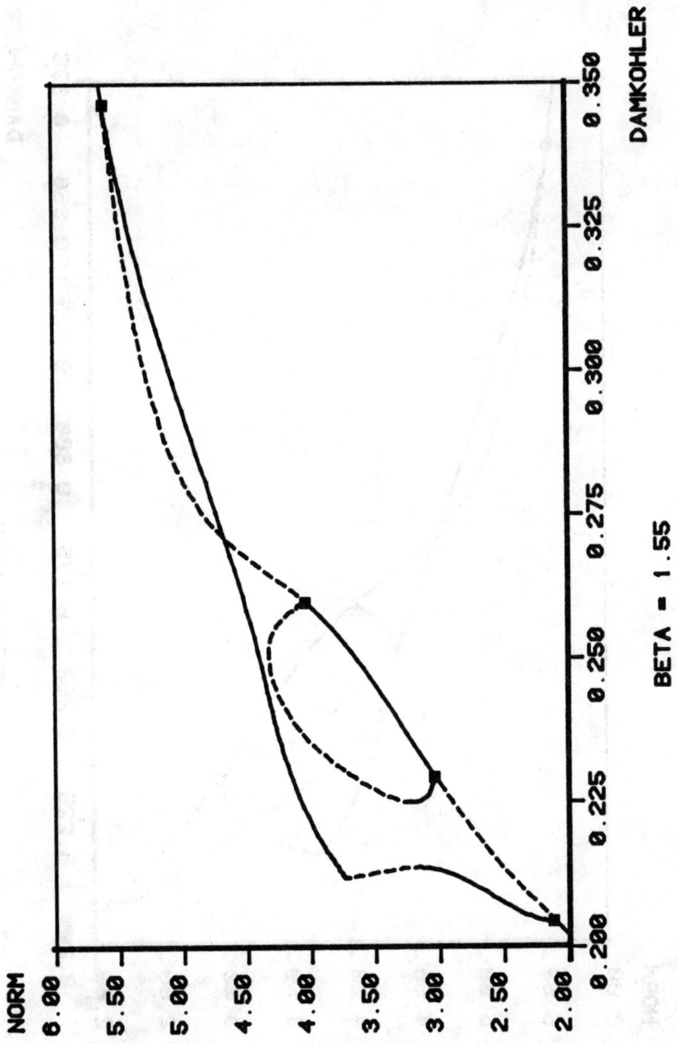
NORM
6.00
5.50
5.00
4.50
4.00
3.50
3.00
2.50
2.00
0.200
0.225
0.250
0.275
0.300
0.325
0.350
DAMKOHLER
BETA = 1.55

FIGURE 11

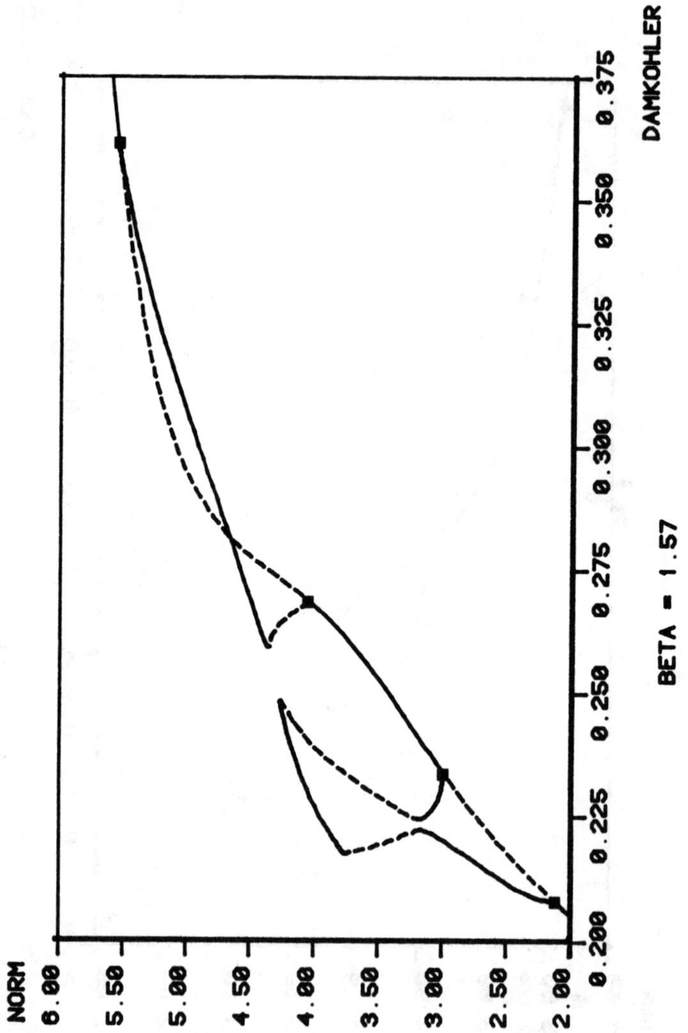
NORM
6.00
5.50
5.00
4.50
4.00
3.50
3.00
2.50
2.00
0.200
0.225
0.250
0.275
0.300
0.325
0.350
0.375
DAMKOHLER
BETA = 1.57

FIGURE 12

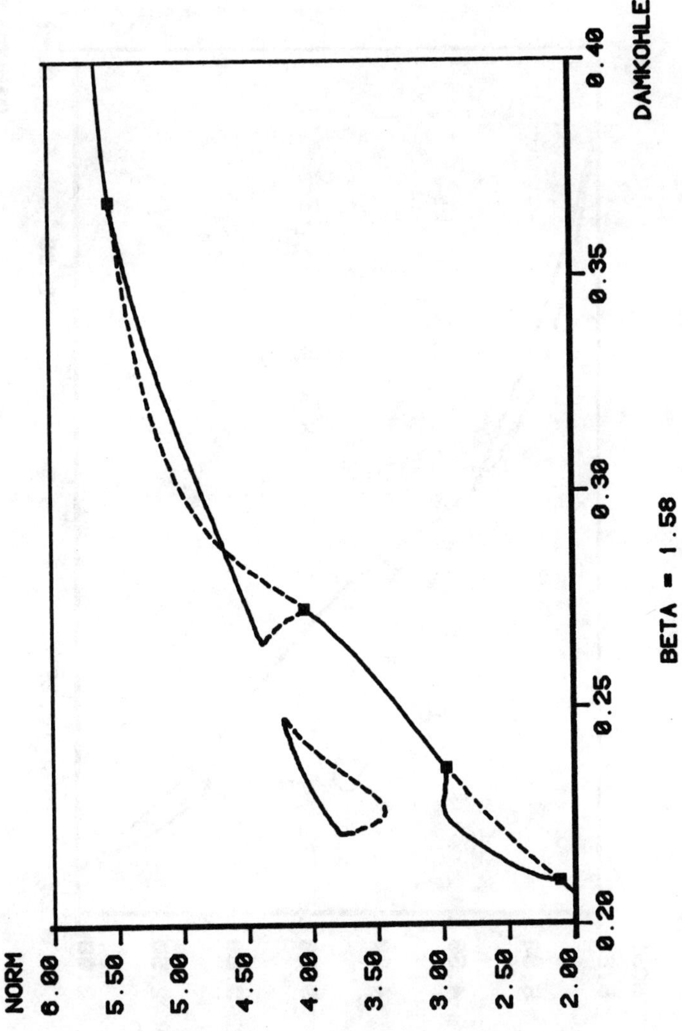
NORM
6.00
5.50
5.00
4.50
4.00
3.50
3.00
2.50
2.00
0.20
0.25
0.30
0.35
0.40
DAMKOHLER
BETA = 1.58

FIGURE 13

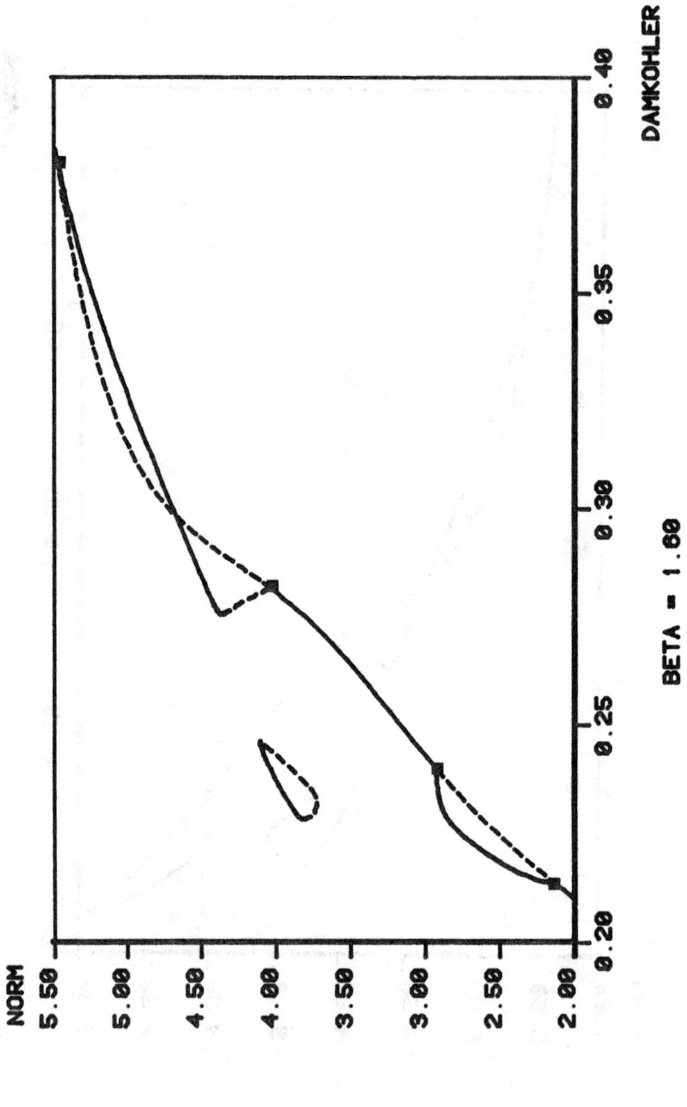
NORM
5.50
5.00
4.50
4.00
3.50
3.00
2.50
2.00
0.20
0.25
0.30
0.35
0.40
DAMKOHLER
BETA = 1.60

FIGURE 14

REFERENCES

[1] D.G. ARONSON, E.J. DOEDEL and H.G. OTHMER, An analytical and numerical study of the bifurcations in a system of linearly-coupled oscillators, manuscript.

[2] U. ASCHER, J. CHRISTIANSEN and R.D. RUSSELL, A collocation solver for mixed order systems of boundary value problems, Math. Comp. 33, 1979, pp. 659-679.

[3] U. ASCHER, J. CHRISTIANSEN and R.D. RUSSELL, Collocation software for boundary value ODE's, ACM TOMS 7, No. 2, 1981, pp. 209-222.

[4] K.-H. BECKER and R. SEYDEL, A Duffing equation with more than 20 branch points, Lecture Notes in Mathematics 878, Springer Verlag, pp. 99-107.

[5] W.-J. BEYN and E.J. DOEDEL, Stability and multiplicity of solutions to discretizations of nonlinear ordinary differential equations, SIAM J. Sci. Stat. Comput. 2, No. 1, 1981, pp. 107-120.

[6] T.N. CHAN, M. Comp. Sci. Thesis, Concordia University, Montreal, 1983.

[7] E.J. DOEDEL and R.F. HEINEMANN, Numerical computation of periodic solution branches and oscillatory dynamics of the stirred tank reactor with A → B → C reactions, Chem. Eng. Sci. 38, No. 9, 1983, pp. 1493-1499.

[8] E.J. DOEDEL, AUTO: A program for the automatic bifurcation analysis of autonomous systems, Cong. Num. 30, 1981, pp. 265-284, (Proc. 10th Manitoba Conf. on Num. Math. and Comput., Winnipeg, Canada).

[9] E.J. DOEDEL and P.C. LEUNG, Numerical techniques for bifurcation problems in delay equations, Cong. Num. 34, 1982, pp. 225-237. (Proc. 11th Manitoba Conf. on Num. Math. and Comput., Winnipeg, Canada).

[10] E.J. DOEDEL and J.P. KERNEVEZ, Mathematical analysis of immobilized enzyme systems, Proc. Int. Conf. on Mathematics in Biology and Medicine, Bari, Italy, July 18— 22, 1983, to appear in Lecture Notes in Biomathematics, Springer Verlag.

[11] E.J. DOEDEL, J.P. KERNEVEZ and D. THOMAS, Bifurcation behaviour of one- and two-compartment enzyme models, manuscript.

[12] M. GOLUBITSKY and W.F. LANGFORD, Classification and unfoldings of degenerate Hopf bifurcations, J. Diff. Eqs. 41, No. 3, 1981, pp. 375-415.

[13] A. GRIEWANK and G.W. REDDIEN, Characterization and computation of generalized turning points, to appear in SIAM J. Numer. Anal.

[14] A. GRIEWANK and G.W. REDDIEN, The calculation of Hopf points by a direct method, to appear in IMA J. Numer. Anal.

[15] K.P. HADELER, Effective computation of periodic orbits and bifurcation diagrams in delay differential equations, Numer. Math., 34, 1980, pp. 457-467.

[16] D.C. HALBE and A.B. POORE, Dynamics of the continuously stirred tank reactor having consecutive reactions A → B → C, MRC Report 2052, University of Wisconsin, 1979.

[17] C. HAYASHI, Selected papers on nonlinear oscillations, Nippon Printing and Publishing Company, Osaka, Japan, 1975.

[18] R.F. HEINEMANN and A.B. POORE, Multiplicity, stability, and oscillatory dynamics of the tubular reactor, Chem. Eng. Sci 36, 1981, pp. 1411.

[19] V. HLAVACEK, M. KUBICEK and K. VISNAK, Multiplicity and stability analysis of a continuous stirred tank reactor with exothermic consecutive reactions A → B → C, Chem. Eng. Sci. 27, 1972, pp. 719-742.

[20] K.F. JENSEN and W.H. RAY, The bifurcation behavior of tubular reactors, Chem. Eng. Sci. 37, No. 2, 1982, pp. 199-222.

[21] A.D. JEPSON, Numerical Hopf bifurcation, Thesis, Part II, California Institute of Technology, Pasadena, Ca., 1981.

[22] A. JEPSON and A. SPENCE, Folds in solutions of two parameter systems and their calculation: Part I, Numerical Analysis Project, Manuscript NA-82-02, Computer Science Department, Stanford University, March 1982.

[23] A. JEPSON, Paths of singular points and their computation, personal communication.

[24] G. JOLY, Analyse des solutions multiples dans les systèmes distribués, Thèse d'Etat, Université de Technologie de Compiègne, France, 1982.

[25] D.V. JORGENSON and RUTHERFORD ARIS, On the dynamics of a stirred tank with consecutive reactions, to appear in Chem. Eng. Sci.

[26] J.P. KEENER, Infinite period bifurcation and global bifurcation branches, SIAM J. Appl. Math. 41, 1981, pp. 127-144.

[27] H.B. KELLER, Numerical solution of bifurcation and nonlinear eigenvalue problems, in: Applications of Bifurcation Theory, P.H. Rabinowitz, ed., Academic Press, 1977, pp. 359-384.

[28] H.B. KELLER, Continuation methods in computational fluid dynamics, in: Numerical and Physical Aspects of Aerodynamic Flows, T. Cebeci, ed., Springer Verlag, 1981.

[29] J. P. KERNEVEZ, Enzyme Mathematics, North-Holland Press, 1980.

[30] J. P. KERNEVEZ, E.J. DOEDEL, M.C. DUBAN, J.F. HERVAGAULT, G. JOLY and D. THOMAS, Spatio-temporal organization in immobilized enzyme systems, Lecture Notes in Biomathematics, Springer Verlag, 1983, pp. 50-75.

[31] W.F. LANGFORD, Periodic and steady state mode interactions lead to tori, SIAM J. Appl. Math. 1979, pp. 22-48.

[32] M. LENTINI and V. PEREYRA, An adaptive finite difference solver for nonlinear two point boundary problems with mild boundary layers, SIAM J. Numer. Anal. 14, 1977, pp. 91-111.

[33] H.D. MITTELMANN and H. WEBER, Numerical methods for bifurcation problems - A survey and classification, in: Bifurcation problems and their numerical solution, ISNM 54, Birkhauser Verlag, 1980, pp. 1-45.

[34] G.W. REDDIEN, Personal communication.

[35] W.C. RHEINBOLDT, Computation of critical boundaries on equilibrium manifolds, SIAM J. Numer. Anal. 19, No. 3, 1982, pp. 653-669.

[36] J. RINZEL and R.N. MILLER, Numerical calculation of stable and unstable periodic solutions to the Hodgkin-Huxley equations, Mathematical Biosciences 49, 1980, pp. 27-59.

[37] R.D. RUSSELL and J. CHRISTIANSEN, Adaptive mesh selection strategies for solving boundary value problems, SIAM J. Numer. Anal. 15, 1978, pp. 59-80.

[38] D.SAUPE, Beschleunigte PL-Kontinuitätsmethoden und periodische Lösungen parametrisierter Differentialgleichungen mit Zeitverzögerung, Dissertation, Universität Bremen, 1982.

[39] D. SAUPE, Global bifurcation of periodic solutions to some autonomous differential delay equations, Report 71, Forschungsschwerpunkt Dynamische Systeme, Universität Bremen, 1982.

[40] I.B. SCHWARTZ, Estimating regions of existence of unstable periodic orbits using computer—based techniques, SIAM J. Numer. Anal. 20, No. 1, 1983, pp. 106-120.

[41] H. WEBER, Numerical solution of Hopf bifurcation problems, Math. Meth. in the Appl. Sc. 2, 1980, pp. 178-190.

NEAR-STOICHIOMETRIC BURNING

ASOK K. SEN* AND G. S. S. LUDFORD**

Abstract. The burning rate of fuel-oxidant mixtures is experimentally observed to be maximum slightly on the fuel-rich side of stoichiometry. The usual explanation is that the flame temperature reaches its maximum on that side because of dissociation of the products. However, there are several unsatisfactory aspects of this explanation. In the present article, a review of a series of papers on the subject, the burning-rate maximum is explained purely on the basis of diffusion of the reactants, i.e. without invoking dissociation of the products. First, a simple model with irreversible reaction, equal molecular masses, and equal specific heats is discussed using dilute-mixture theory. This is followed by a generalization to unequal molecular masses (and, later, unequal specific heats). We finally consider a reversible reaction to show that, while dissociation explains the fuel-rich maximum of the flame temperature, it has only a subsidiary effect on the burning-rate maximum. Results for non-dilute mixtures are also discussed.

NOMENCLATURE

a,b,c	parameters appearing in (1.5), (1.3) and (1.2)
c_p	specific heat of mixture at constant pressure (1.8)
D	Damköhler number (2.3)
D_{ij}	binary diffusion coefficient of species i and j (1.8)
k	equilibrium constant (3.4)
K,L	Lewis numbers of oxidant, fuel (1.6)
m	molecular mass (2.17)
M	burning rate
N	mass fraction of inert
R	Lewis number of product (3.2)
T	temperature

*Department of Mathematical Sciences, Purdue School of Science at Indianapolis, Indianapolis, IN 46223.
**Center for Applied Mathematics, Cornell University, Ithaca NY 14853.
Research supported by the U.S. Army Research Office.

x	dimensionless distance (unit $\lambda/c_p M$)
X,Y,Z	mass fractions of oxidant, fuel, product
α,β	stoichiometric parameters (2.16)
γ	stoichiometric mass ratio (2.17)
θ	activation temperature (2.2)
λ	thermal conductivity of mixture (1.8)
ρ	density of mixture (1.8)
ν	stoichiometric coefficients (2.14)
Ω	reaction rate (2.2)

Subscripts

e	effective value
F	fuel
O	oxidant
P	product
*	conditions at the flame
$-\infty$	fresh mixture

1. Introduction. The subject of near-stoichiometric burning has been of considerable interest for many years. It is experimentally known that the flame temperature and the burning rate of a combustible mixture reach their maximum values when the unburnt mixture is slightly fuel-rich. The burning-rate maximum is usually explained by the fact that the flame temperature itself attains its maximum slightly on the fuel-rich side of stoichiometry, due to dissociation of the combustion products. As pointed out by Sen & Ludford [1]*, there are unsatisfactory aspects of such an explanation: the burning rate depends not only on the flame temperature but also on the concentrations of the reactants, especially near stoichiometry; and the explanation does not apply to flames in which dissociation may be negligible, e.g. cool flames.

SL-1 explained this near-stoichiometric behavior of the maximum burning rate purely on the basis of diffusion of the reactants (i.e. without invoking dissociation of the products). They considered a one-step irreversible reaction

$$F + O \rightarrow P, \tag{1.1}$$

with equal molecular masses for the reactants and equal specific heats for all species. Here F, O, P represent fuel, oxidant and product molecules respectively. The reaction (1.1) is assumed to take place in a large amount of an inert species N.

Experiments are generally performed under two conditions: either the concentration of the inert species in the unburnt mixture is kept fixed so that

*Hereafter referred to as SL-1.

(1.2) case I: $X_{-\infty} + Y_{-\infty} = 1 - N_{-\infty} = c(\text{const.})$

holds; or the inert-oxidant ratio in the unburnt mixture is fixed, i.e.

(1.3) $N_{-\infty}/X_{-\infty} = b(\text{const.})$

and we have

(1.4) case II: $(b+1)\, X_{-\infty} + Y_{-\infty} = 1.$

The two cases can be combined by writing

(1.5) $N_{-\infty} = a + b\, X_{-\infty}$ with $a = 1 - c,$

so that the conditions $a \neq 0,\ b = 0$ refer to case I and $a = 0,\ b \neq 0$ refer to case II.

Using a dilute mixture theory, SL-1 show that for the scheme (1.1) the maximum flame temperature in both cases I and II occurs exactly at stoichiometry. However, the burning rate is found to reach its maximum slightly on the fuel-rich side of stoichiometry, provided

(1.6) $K > 1,\quad L < 1$ for case I,

(1.7) $K > (b+2)/2(b+1),\quad L < 1 + b/2$ for case II.

The Lewis numbers K and L of the oxidant and the fuel have the definitions

(1.8) $K = \lambda/\rho D_{ON} c_p,\quad L = \lambda/\rho D_{FN} c_p.$

The D_{ij}'s are the binary diffusion coefficients for species i and j, and λ and ρ are the thermal conductivity and density of the mixture respectively. The conditions (1.6) and (1.7) are almost always met in practice. Diffusion of the reactants may therefore be considered a satisfactory explanation for the fuel-rich maximum of the burning rate.

Although the simple model used by SL-1 is sufficient to explain the maximum burning rate, it is overly restrictive in the sense that (i) the reaction (1.1) is too simple, (ii) equality of molecular masses and specific heats is unrealistic, and (iii) the model does not apply to the non-dilute mixtures that are invariably encountered in practice (e.g. fuel-air mixtures). In subsequent papers (Ludford & Sen [2], [3]; Sen & Ludford [4], [5], [6]), these deficiencies were removed to show that in all cases of practical interest the maximum burning rate occurs on the fuel-rich side purely as the result of diffusion of the reactants. They have also introduced dissociation of the products to demonstrate that, while this explains the fuel-rich maximum of flame temperature, as is usually believed, it has only a subsidiary effect on the location of the maximum burning rate. These matters are discussed in the following sections.

2. Dilute-Mixture Theory with No Dissociation. The chemical reactions in most combustion processes are usually very complicated, consisting of a large number of elementary steps occuring simultaneously or in succession. However, for analytical studies it is sufficient to adopt a global reaction-rate theory, i.e. replace the overall combustion process by a single one-step irreversible reaction such as (1.1). Furthermore, the reactive mixture is very often assumed to be dilute, so as to have the simplicity of Fick's law for the diffusion of species (Buckmaster & Ludford [7]). We now discuss the model reaction (1.1) using a dilute-mixture theory.

2A. Equal Molecular Masses, Equal Specific Heats. The conservation equations in one space dimension can be written

(2.1) $$T' - T'' = 2(K^{-1} X'' - X') = 2(L^{-1}Y'' - Y') = \Omega$$

with

(2.2) $$\Omega = \Lambda XY \exp(-\theta/T)$$

and

(2.3) $$\Lambda = DM^{-2}.$$

Here a prime denotes d/dx. The upstream boundary conditions are

(2.4) $$x = -\infty: \quad T = T_{-\infty}, \; X = X_{-\infty}, \; Y = Y_{-\infty}.$$

In writing down equations (2.1) it has been assumed that the reactants F and O have the same molecular masses and all species (including the diluent N) have the same specific heats. Furthermore, the mixture is considered to be highly dilute in the inert species N so that Fick's law for binary diffusion applies.

If the unburnt mixture is fuel-rich ($Y_{-\infty} > X_{-\infty}$), the oxidant is depleted at the flame, i.e.

(2.5) $$X_* = 0.$$

Then equations (2.1b,c) show that the fuel fraction at the flame has the value

(2.6) $$Y_* = Y_{-\infty} - X_{-\infty}.$$

This is nothing more than conservation of fuel and oxidant atoms in in the reaction (1.1). Similarly, equations (2.1a,b) yield the flame temperature

(2.7) $$T_* = T_{-\infty} + 2X_{-\infty},$$

a result that simply expresses conservation of total enthalpy.

Using the relations (2.6), (2.7), (1.2) and (1.4), it can be easily shown that T_* becomes maximum for $Y_* = 0$, i.e. at exact stoichiometry. This simple model therefore shows that, in the absence of dissociation, the maximum flame temperature always occurs for a stoichiometric mixture.

As indicated above, this conclusion could also be reached with purely thermodynamic considerations, i.e. without using the equations (2.1). In determining the burning rate, however, these equations must be used. SL-1 analyzed them in the limit of large activation energy ($\theta \to \infty$), when the reaction is concentrated in a sheet outside which there is negligible reaction. (The method of activation-energy asymptotics consists in finding expansions in the outer regions and matching them with expansions in the flame-sheet layer.) In this way SL-1 show that, when the fresh mixture is away from stoichiometry, i.e. $Y_* = O(1)$, the burning rate is given by

(2.8) $$M^2 = \frac{DKY_*T_*^4}{4\theta^2 X_{-\infty}^2} \exp(-\theta/T_*).$$

This formula predicts (incorrectly) that the burning rate goes to zero as $Y_* \to 0$. However, when the fresh mixture is close to stoichiometry, i.e.

(2.9) $$Y_* = \theta^{-1} y_* \quad \text{with} \quad y_* = O(1),$$

the analysis breaks down and must be modified. The fuel-rich burning rate is then given by

(2.10) $$M^2 = \frac{DKT_*^4}{4\theta^3 X_{-\infty}^2} (y_* + LT_*^2)\exp(-\theta/T_*).$$

An analogous formula can easily be derived for a fuel-lean mixture near stoichiometry.

Using this expressions for M shows that M reaches it maximum value, slightly on the fuel-rich side of stoichiometry, only if the Lewis numbers K and L satisfy the inequalities (1.6) for case I and (1.7) for case II. The maximum is located at

(2.11) $$y_* = (1 + b/2 - L)T_{*0}^2,$$

where

(2.12) $$T_{*0} = T_{-\infty} + 2c/(b+2)$$

is the leading approximation to the flame temperature. Recall that for case I, $b = 0$, $c \neq 1$ and for case II, $b \neq 0$, $c = 1$.

The full results for case I are shown on a K,L-plane in figure 1, which is reproduced from SL-1. For case II, K and L should be replaced by $[2(b+1)/(b+2)]K$ and $2L/(b+2)$ respectively, in accordance with (1.7).

To test the validity of these results consider, for instance, the burning of hydrocarbon-air mixtures, for which $b = 3.29$ (case II). The inequalities (1.7) become

(2.13) $$K > 0.62, \quad L < 2.64,$$

conditions that are always satisfied, since the Lewis number (K) of oxygen is slightly greater than one and the Lewis numbers (L) of hydrocarbons are less than 2.64. For case I, consider hydrocarbon-oxygen-nitrogen mixtures in which the concentration of nitrogen is kept fixed. Our theory requires the inequalities (1.6) to hold if the burning-rate maximum is to be on the fuel-rich side. Light hydrocarbons satisfy these conditions since their Lewis numbers (L) are usually less than one, while (as noted above) the Lewis number of oxygen is slightly greater than one. Heavier hydrocarbons on the other hand have $L > 1$, yet experiments show that the burning rate still has a fuel-rich maximum. The discrepancy is due to the assumption of equal molecular masses. In the following subsection we show that, when the effect of different molecular masses is taken into account, the maximum burning rate is predicted to be on the fuel-rich side for all mixtures in case I (and in case II, of course).

2B. Unequal Molecular Masses and Specific Heats. In studying the effect of unequal molecular masses on the maximum flame temperature and burning rate in the absence of dissociation, Ludford & Sen [2] considered a more general reaction

(2.14) $$\nu_F F + \nu_O O \rightarrow \nu_P P.$$

The governing equations for this scheme are

(2.15) $$T' - T'' = \alpha^{-1}(K^{-1}X''-X') = \beta^{-1}(L^{-1}Y''-Y') = \Omega,$$

which are subject to the upstream boundary conditions (2.4). Here

(2.16) $$\alpha = \gamma/(\gamma+1) \quad \text{and} \quad \beta = 1/(\gamma+1),$$

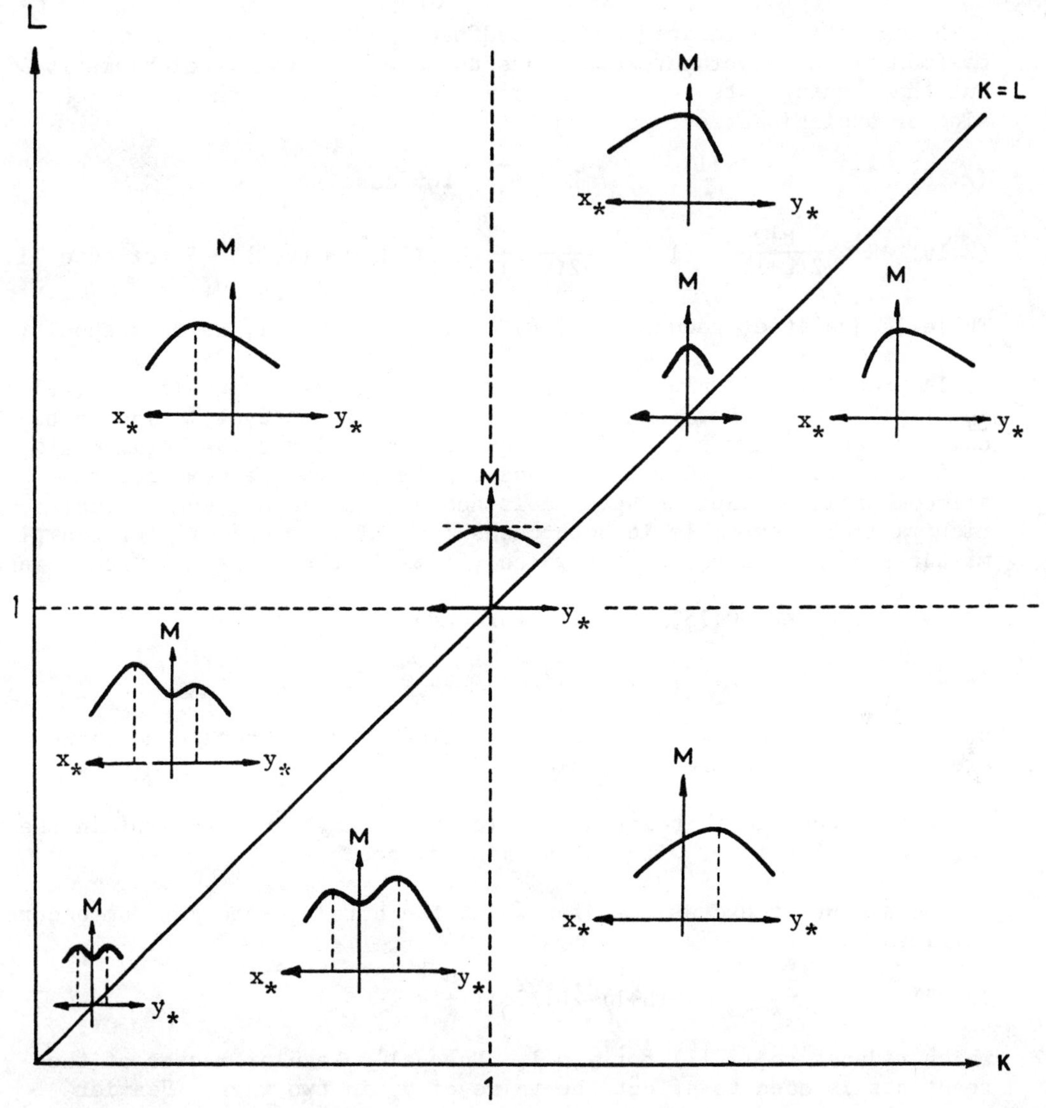

Figure 1. K, L-plane with shape of burning-rate curves in the various regions. Distance from stoichiometry on the fuel-rich side is measured by the y_* of equation (2.9); distance on the fuel-lean side is measured by the corresponding x_*.

where

$$(2.17) \qquad \gamma = \gamma_0 m_0/\nu_F m_F,$$

is the so-called stoichiometric mass ratio.

Using this generalized model, Ludford and Sen showed that the maximum flame temperature continues to occur at exact stoichiometry, but the burning rate reaches its maximum slightly on the fuel-rich side of stoichiometry, provided

$$(2.18) \qquad K > \frac{\gamma+1}{2\gamma}, \qquad L < \frac{\gamma+1}{2} \text{ for case I,}$$

$$(2.19) \quad K > \frac{b+2}{2(b+1)} - (1 - \frac{1}{\gamma}) \frac{1}{2(b+1)}, \; L < (1+\frac{b}{2}) + (\gamma-1) \frac{b+1}{2} \text{ for case II.}$$

These inequalities reduce to (1.6), (1.7) when $\gamma = 1$, as they should.

In practice γ is always greater than 1, so that inequalities (2.18) and (2.19) are weaker than (1.6) and (1.7), respectively. The vertical divider in figure 1 is displaced to the left and the horizontal divider is moved up as γ is increased from unity. We now find that all combustible mixtures have Lewis numbers (K,L) that make a fuel-rich maximum inevitable in both cases I and II. Consider, for example, mixtures of hydrocarbons (for which γ lies in the range 2-4) requiring

$$(2.20) \qquad K > 0.75, \quad L < 1.5 \text{ for case I,}$$

$$(2.21) \qquad K > 0.56, \quad L < 4.79 \text{ for case II;}$$

here we have used the most stringent value $\gamma = 2$. These conditions are met by all mixtures we could find.

We may conclude that diffusion alone is always able to explain the burning-rate maximum, without invoking dissociation.

The amount of excess fuel for which the burning-rate maximum occurs is given by

$$(2.22) \qquad y_* = [1+\gamma(b+1)-2L]T_{*0}^2/(\gamma+1),$$

which reduces to (2.11) for $\gamma = 1$. Unequal molecular masses of the reactants is seen to affect the value of y_* in two ways. Heavier fuels have higher Lewis numbers (L), due to their lower rates of diffusion; on the other hand, such fuels have a higher stoichiometric mass ratio (γ). These two effects compete in the formula (22) and it is not clear which one dominates. However, for actual mixtures, the effect of L usually outweights the γ-effect. Thus, with heavier fuels one would expect the burning-rate maximum to move toward stoichiometry and this trend has been generally observed with hydrocarbon-air mixtures (see Lichty [8]). On the other hand, mixtures containing hydrogen have a high value of the stoichometric mass ratio ($\gamma = 8$) and a low value of L (approximately .29), so that the two effects reinforce each other yielding a maximum burning rate further away from stoichiometry, in agreement with experimental results.

The effect of unequal specific heats is minor and will not be discussed here. The interested reader is referred to Ludford & Sen [2].

3. Dilute Mixtures with Dissociation. In our discussion so far, dissociation of the products was ignored. We found that the flame temperature occurs exactly at stoichiometry and the burning-rate maximum on the fuel-rich side can be explained by diffusion of the reactants alone. In this section, dissociation of the products is included to show that it indeed shifts the maximum flame temperature to the fuel-rich side of stoichiometry; but there is only a subsidiary effect on the location of the maximum burning rate.

3A. The Simple Model with Dissociation. The simplest way to incorporate dissociation into the previous models is to include the reverse reaction; we consider the process

$$F + O \rightleftharpoons P \tag{3.1}$$

with equal molecular masses of the reactants and equal specific heats for all species. The equations (2.1) are now augmented by an equation for the product, namely,

$$R^{-1}Z'' - Z' = -\Omega. \tag{3.2}$$

Likewise, to the upstream boundary conditions (2.4) is added

$$x = -\infty:\ Z = 0 \tag{3.3}$$

(assuming that no product species is present in the unburnt mixture). The reaction term Ω is also modified to

$$\Omega = \Lambda(XY-kZ)\exp(-\theta/T). \tag{3.4}$$

By eliminating the reaction term Ω from equations (2.1) and (3.2), and integrating the result, we find

$$T_* = T_{-\infty} + Z_*,\ Y_* = Y_{-\infty} - X_{-\infty} + X_*,\ Z_* = 2(X_{-\infty} - X_*). \tag{3.5}$$

These relations can be alternatively derived from the conservation of fuel and oxidant atoms and enthalpy (cf. section 2). The equilibrium condition

$$k(T_*) = X_*Y_*/Z_* \tag{3.6}$$

(which in general depends on the flame temperature) provides the fourth relation between X_*, Y_*, Z_* and T_*. The variation of k with temperature is known from experiments. Calculation of the flame temperature or of any of the equilibrium products from equations (3.5) and (3.6) clearly involves an iterative procedure in general.

Sen & Ludford [4] have shown that when the dissociation is weak (as is invariably the case in practice), an asymptotic approximation of the flame temperature can be found analytically, without using an iterative procedure. Consider, for definiteness, a slightly fuel-rich mixture with the excess fuel given by

$$Y_{-\infty} - X_{-\infty} = \theta^{-1}y_*, \quad y_* = O(1); \tag{3.7}$$

and let the dissociation be weak in the sense that

$$X_* = \theta^{-1}\hat{X}_*, \quad \hat{X}_* = O(1), \tag{3.8}$$

i.e. only an $O(\theta^{-1})$ amount of oxidant is produced as a result of the dissociation. Then the flame temperature is

$$T_* = T_{*0} - \theta^{-1}T_{*1}, \tag{3.9}$$

where T_{*0} has the value (2.7) with $X_{-\infty} = c/(b+2)$ and

$$T_{*1} = 2[\hat{X}_* + y_*/(b+2)]. \tag{3.10}$$

In this result, which comes from the relations (3.5), $\hat{X}_*$ is to be expressed in terms of y_* by means of the remaining relation (3.6). We find the quadratic equation

$$\hat{X}_*(y_* + \hat{X}_*) = 2c\hat{k}(T_{*0})/(b+2) \tag{3.11}$$

for $\hat{X}_*$, when we write

$$k(T_*) = \theta^{-2}\hat{k}(T_*) = \theta^{-2}[\hat{k}(T_{*0}) + O(\theta^{-1})]. \tag{3.12}$$

Equations (3.10) and (3.11) show that T_{*1} has a minimum at

$$y_* = b\left[\frac{2c\hat{k}(T_{*0})}{(b+1)(b+2)}\right]^{\frac{1}{2}}. \tag{3.13}$$

This result clearly shows that in case II ($b \neq 0$) the maximum flame temperature occurs slightly on the fuel-rich side; but that for case I ($b = 0$) it does not. Taking into account unequal molecular masses changes this last result as we shall see in the following subsection.

A similar analysis shows that there is no maximum on the fuel-lean side of stoichiometry.

In describing the results for maximum burning rate it is convenient to treat cases I and II separately. We will confine our discussion to that region of the K,L-plane (see figure 1) where, in the absence of dissociation, there is a single maximum of M and it lies on the fuel-rich side of stoichiometry. The purpose is to see how this maximum is affected when dissociation is admitted, i.e. as the equilibrium parameter $\hat{k}$ is increased from zero for fixed values of K and L.

Consider case I first; the results are shown in figures 2 and 3. These figures have been drawn for values of K and L that lie in the region of interest. When $\hat{k} = 0$ (no dissociation) the maximum occurs at $y_* = (1-L)T_{*0}^2$, in accordance with the result (2.11). As $\hat{k}$ increases, this maximum moves back toward or farther away from stoichiometry according as $K+L < 2$ (figure 2) or $K+L > 2$ (Figure 3), finally approaching $y_* = (K-L)T_{*0}^2/(K+L)$ as $\hat{k} \to \infty$. For $K+L = 2$, dissociation has no effect on the burning-rate maximum, and we note that there are many mixtures approximately satisfying this condition: K is slightly greater than one and L is slightly less than one.

The results for case II are given in figure 4. Here the maximum continually moves away from stoichiometry as k increases, but no new maximum appears.

3B. <u>The Generalized Model with Dissociation</u>. We now consider the reversible reaction

$$\nu_F F + \nu_0 0 \rightleftharpoons \nu_P P \tag{3.14}$$

with unequal molecular masses and specific heats. Here the equations (2.15) apply but with the reaction term (3.4).

Sen & Ludford [6] show that, for a slightly fuel-rich mixture, the flame temperature T_* can still be written in the form (3.9) but now with

$$T_{*1} = \frac{\gamma+1}{1+\gamma(b+1)} y_* + \frac{\gamma+1}{\gamma} \hat{X}_* \tag{3.15}$$

and

$$\hat{X}_*(y_*+\gamma^{-1}\hat{X}_*) = (\gamma+1)c\hat{k}(T_{*0})/[1+\gamma(b+1)] \tag{3.16}$$

replacing equations (3.10) and (3.11), respectively. It follows that the flame temperature reaches a maximum on the fuel-rich side only if

$$\gamma(b+1) > 1. \tag{3.17}$$

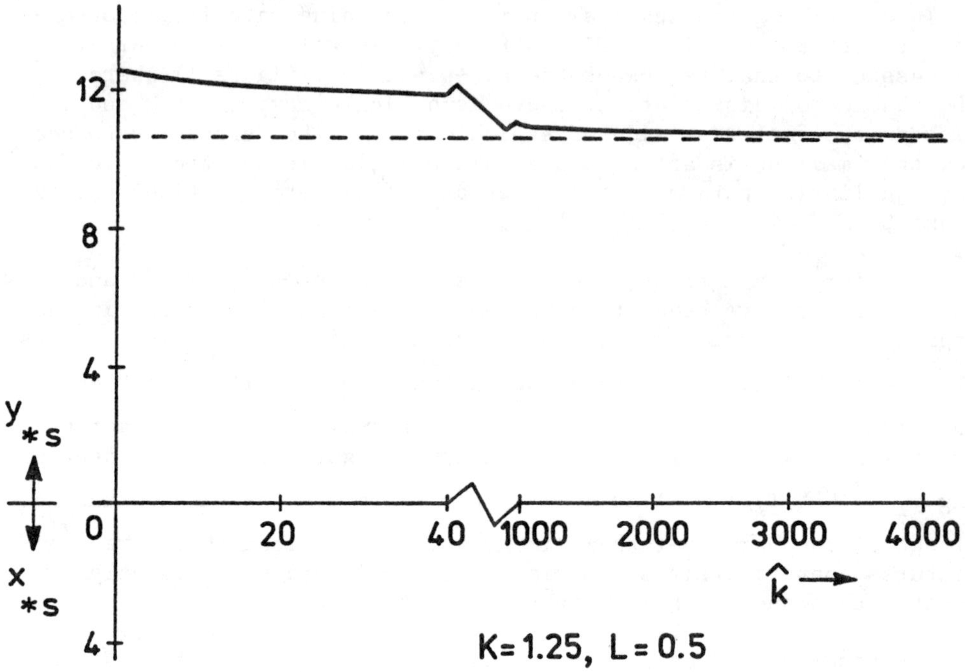

Figure 2. Case I: effect of dissociation on the fuel-rich maximum for K = 1.25, L = 0.5 according to section 3A.

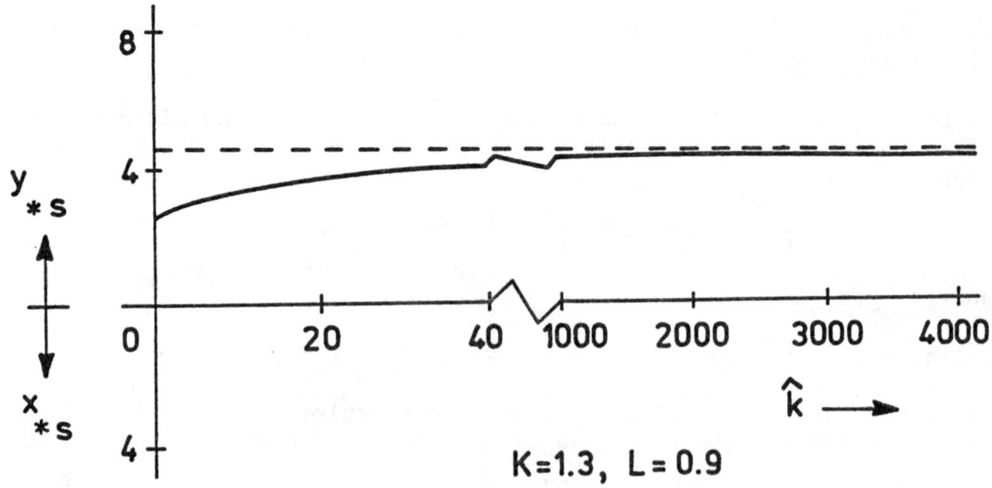

Figure 3. Case I: effect of dissociation on the fuel-rich maximum for K = 1.3, L = 0.9 according to section 3A.

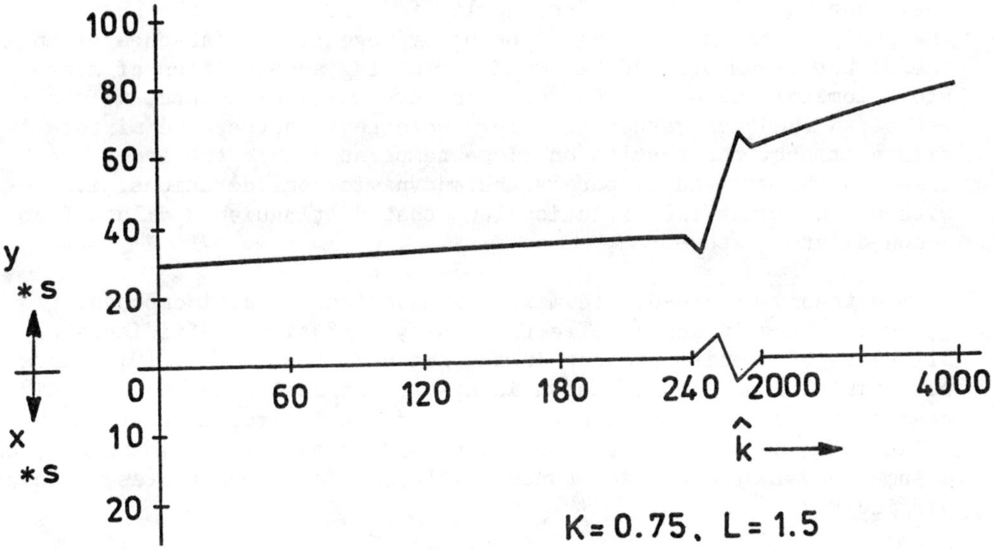

Figure 4. Case II: effect of dissociation on the fuel-rich maximum for K = 0.75, L = 1.5 according to section 3A.

For

$$(3.18) \qquad \gamma(b+1) < 1,$$

the maximum is on the fuel-lean side. If we recall that γ ranges from 2 to 4 for mixtures of hydrocarbons with oxygen or air, then the inequality (3.17) is satisfied whether or not b equals zero. Thus all hydrocarbon-oxygen-inert (case I) and hydrocarbon-air (case II) mixtures should attain their maximum flame temperature slightly on the fuel-rich side of stoichiometry, and this is confirmed by experimental observation (Khitrin [9]). Other fuels such as hydrogen behave in the same manner, a notable exception being carbon monoxide in case I: the inequality (3.18) is satisfied instead of (3.17), but experiments show that the flame temperature is a maximum when the fresh mixture is slightly fuel-rich.

The results for the maximum burning rate remain qualitatively the same as for $\gamma = 1$ (see section 3A) except that now it moves steadily away from stoichiometry as $\hat{k}$ increases in both cases I and II. The conclusion is that dissociation merely shifts the fuel-rich maximum of the burning rate further away from stoichiometry.

4. Results for Non-Dilute Mixtures. It should be emphasized that the theories of the previous sections assume that the combustible mixture is very dilute in the reactants. At first sight it would seem that such theories do not apply, for example, to the burning of the fuels in air or the oxidation of hydrogen in the absence of an inert; but arguments can be found for making an exception of near-stoichiometric burning. Rather than give these arguments, we shall establish the fact mathematically. Note that whether the mixture is dilute or not, the results on flame temperature are the same, since these can be derived by purely thermodynamic considerations, i.e. without involving the diffusion laws that distinguish a dilute from a non-dilute mixture.

The theory of non-dilute mixtures is based on multicomponent diffusion laws, which involve the binary diffusion coefficients of all pairs of species present in the mixture. As Clarke [10] points out, "multicomponent diffusion is awkward to deal with and for this reason Fick's law is frequently used outside its realm of strict validity". However, if the molecular masses of all species are assumed equal, the multicomponent diffusion laws do not present any difficulty.

Sen & Ludford [5] first derived the burning rate of a non-dilute mixture in the absence of dissociation. Their main result is that a maximum exists slightly on the fuel-rich side of stoichiometry if the inequalities (1.6) and (1.7), namely

$$K_e > (b+2)/2(b+1), \quad L_e < 1+b/2 \tag{4.1}$$

(with b = 0 for case I), hold for the "effective" Lewis numbers

$$K_e = L_{ON} + 2X_{-\infty}(L_{OP}-L_{ON}), \quad L_e = L_{FN} + 2X_{-\infty}(L_{FP}-L_{FN}), \tag{4.2}$$

respectively, where

$$X_{-\infty} = (1-a)/(b+2). \tag{4.3}$$

Here the definitions (1.8) apply to $K = L_{ON}$ and $L = L_{FN}$; the remaining Lewis numbers are found from analogous formulas, involving the other binary diffusion coefficients. Surprisingly enough, these effective Lewis numbers do not depend on L_{OF}: the diffusion of one reactant with respect to the other has no effect in near-stoichiometric mixtures. Note that when $L_{OP} = L_{ON}$ and $L_{FP} = L_{FN}$, as is very nearly so for many mixtures, K_e and L_e become identical to K and L, i.e. the mixture behaves as if it were dilute. (Otherwise new phenomena of doubtful practical significance result; these are worked out in detail by Sen & Ludford [5]).

For non-dilute mixtures containing no inert, the effective Lewis numbers are

$$K_e = L_{OP}, \quad L_e = L_{FP}, \tag{4.4}$$

so that the mixture behaves exactly like a dilute mixture with its inert species replaced by the product.

Ludford & Sen [3] then included dissociation of the product in this equal-molecular-mass model of non-dilute mixtures and found that the dilute-mixture results of subsection 3A still hold, provided K and L are replaced by the same effective Lewis numbers (4.2).

5. Concluding Remarks. The series of papers reviewed here shows that diffusion of the reactants is capable by itself of producing the observed fuel-rich maximum in the burning rate of combustible mixtures. Dissociation, the mechanism invoked by the simpler chemical-kinetic explanations, is seen to have a subsidiary effect, simply changing the location of the maximum established by diffusion. Any explanation of the burning-rate maximum based soley on chemical kinetics, however complex, is incomplete and misleading; however, the precise location of the maximum can be expected to depend on the chemical kinetics.

By contrast, the fuel-rich maximum of the flame temperature, at which the simpler explanations are actually aimed, is due to dissociation. The theory gives an explicit analytical demonstration of this fact.

REFERENCES

[1] A.K. SEN & G.S.S. LUDFORD, The near-stoichiometric behavior of combustible mixtures. Part I: diffusion of the reactants, Combust. Sci. Technol. 21 (1979), pp. 15-23.

[2] G.S.S. LUDFORD & A.K. SEN, Burning rate maximum of a plane pre-mixed flame, Progr. Aeronaut. Astronaut. 76 (1981), pp. 427-436.

[3] G.S.S. LUDFORD & A.K. SEN, The effect of dissociation on the near-stoichiometric burning of non-dilute mixtures, Progr. Aeronaut. Astronaut., to appear.

[4] A.K. SEN & G.S.S. LUDFORD, The near-stoichiometric behavior of combustible mixtures. Part II: dissociation of the products, Combust. Sci. Technol. 26 (1981), pp. 183-191.

[5] A.K. SEN & G.S.S. LUDFORD, Effects of mass diffusion on the burning rate of non-dilute mixtures, 18th Symp. (int.) Combust. (The Combustion Institute 1981), pp. 417-424.

[6] A.K. SEN & G.S.S. LUDFORD, Maximum flame temperature and burning rate of combustible mixtures, 19th Symp. (int.) Combust. (The Combustion Institute 1983), pp. 267-274.

[7] J.D. BUCKMASTER & G.S.S. LUDFORD, Lectures on mathematical combustion, SIAM Publications, Philadelphia, 1983.

[8] L.C. LICHTY, Combustion engine processes, McGraw Hill, New York, 1967.

[9] L.N. KHITRIN, The physics of combustion and explosion, U.S. Department of Commerce, Washington, D.C., 1962.

[10] J.F. CLARKE, Parameter perturbations in flame theory, Progr. Aerospace Sci. 16 (1975), pp. 3-29.

PART III: MATHEMATICAL MODELLING IN HYDROCARBON RECOVERY

A COLLOCATION MODEL OF COMPOSITIONAL OIL RESERVOIR FLOWS

MYRON B. ALLEN*

Abstract. Finite element collocation offers many of the advantages of Galerkin's method with the potential for considerable computational savings. However, difficulties with spurious oscillations and convergence problems have discouraged the use of collocation in simulating multiphase porous-media flows. A one-dimensional simulator demonstrates the applicability of collocation to the types of compositional reservoir flows occurring, for example, in miscible gas floods. For simplicity in developing the method, it is useful to consider a model that treats two-phase flows with interphase mass transfer.

The resulting simulator uses the Peng-Robinson equation of state to predict fluid-phase thermodynamics. The code solves the species flow equations using an implicit pressure - explicit composition formulation based on upstream collocation in spaces of Hermite cubics. This upstream-biased scheme is analogous to upstream weighting in finite-difference theory.

In addition, an explicit geometric representation of the saturation-pressure surface furnishes an efficient, numerically reliable method for computing phase equilibria. This representation eliminates the need for solving nonlinear equal-fugacity constraints during simulation, yet it preserves the thermodynamic consistency of the standard equation-of-state approaches. The scheme renders phase compositions and saturations with order-of-magnitude or greater savings in CPU time.

1. Introduction. Several technologies for the enhanced recovery of crude oil depend for their effectiveness on mechanisms that reduce capillary forces trapping oil in the voids of porous reservoir rocks. Prominent among these technologies is miscible gas flooding, in which an injected vapor mixes with the oil and continually exchanges molecular species across the vapor-liquid phase boundary. When successful, this process generates a transition zone of fluid in which the interfacial tension between vapor and liquid is very small. As this low-tension mixture moves from injection well to producing well it

*Department of Mathematics, University of Wyoming, Laramie, WY 82071

mobilizes the trapped oil and drives it toward producing wells.

Mathematical models of this class of technologies must account for the simultaneous flow of several fluid phases in a porous medium and also for the effects of interphase mass transfer on the fluid-rock system. Such models are called compositional models. To date the petroleum industry has relied largely on the method of finite differences to generate discrete approximations to the partial differential equations governing compositional flows. This paper presents an alternative method, finite-element collocation. In particular, we shall examine a technique called upstream collocation, which is analogous to the commonly used upstream-weighted difference schemes. This technique guarantees convergence of the discrete analogs in highly convective, nonlinear flows. The applicability of collocation to the simplified problems discussed here suggests that the method may ultimately prove useful in industrial-scale simulators.

The actual numerics appear in Sections 4 and 5 of this paper. Before discussing them, we shall review the basic continuum mechanics of an oil reservoir and briefly comment on their implications for numerical methods.

2. Basic Reservoir Mechanics. The equations governing multicomponent, multiphase flows in porous media arise from the constituent mass and momentum balance laws of continuum mixture theory, along with thermodynamic constraints on fluid compositions and pressures. For simplicity, let us assume that the reservoir consists of three phases, V, L, and R (vapor, liquid, and rock, respectively), and that there are N molecular species shared by the two fluid phases together with an (N+1)-st species constituting the rock matrix. Associated with each species i in each phase α is an intrinsic molar density ρ_i^α having dimensions [moles of i in phase α/volume of α], and associated with each phase α is a dimensionless volume fraction ϕ_α having dimensions [volume of α/volume of mixture]. The set of volume fractions obeys $\phi_V + \phi_L + \phi_R = 1$.

The molar densities and volume fractions give rise to several purely logical restrictions on certain derived quantities describing the mixture. Thus we have, for example, the porosity $\phi = \phi_V + \phi_L$ and hence the saturations

$$S_V = \phi_V/\phi, \quad S_L = \phi_L/\phi$$

of the fluid phases, which must obey the restriction

$$S_V + S_L = 1 \tag{2-1}$$

Also, the intrinsic molar density of any phase α is $\rho^\alpha = \rho_1^\alpha + \cdots + \rho_{N+1}^\alpha$, and using this we can define the mole fraction of species i in phase α, $\omega_i^\alpha = \rho_i^\alpha/\rho^\alpha$. Let us also define the overall molar density of the fluids,

$$\rho = \phi(S_V\rho^V + S_L\rho^L)$$

and thus the overall mole fraction of species i in the fluids,

$$\omega_i = \phi(S_V\rho^V\omega_i^V + S_L\rho^L\omega_i^L)/\rho$$

These new variables obey the restrictions

$$\text{(2-2)} \qquad \sum_i \omega_i^V = \sum_i \omega_i^L = \sum_i \omega_i = 1$$

The set of constituents, viewed as a collection of ordered pairs (i,α), must satisfy the mass balance,

$$\text{(2-3)} \qquad \frac{d}{dt}\left(\sum_i \sum_\alpha \int_\Gamma \rho_i^\alpha \, dv\right) = 0$$

where Γ can be any spatial volume occupied by material from the reservoir Ω. A standard argument (Eringen and Ingram [11]) reduces this equation to a set of point balances

$$\text{(2-4)} \qquad \frac{\partial \rho_i^\alpha}{\partial t} + \nabla\cdot(\rho_i^\alpha \underset{\sim}{v}_i^\alpha) = \hat{\rho}_i^\alpha$$

on any domain where the variables ρ_i^α and $\underset{\sim}{v}_i^\alpha$ are sufficiently smooth. In this equation $\underset{\sim}{v}_i^\alpha$ is the velocity of constituent (i,α), and $\hat{\rho}_i^\alpha$ is the molar rate of exchange of mass into constituent (i,α) from other constituents. For the individual constituent balances (2-4) to be consistent with the global mass balance (2-3), the exchange terms must satisfy

$$\sum_i \sum_\alpha \hat{\rho}_i^\alpha = 0$$

For simplicity, let us restrict attention to mixtures in which no homogeneous chemical reactions occur, so that $\hat{\rho}_i^V + \hat{\rho}_i^L + \hat{\rho}_i^R = 0$ for all i, and in which the rock phase shares no species with the fluids; in symbols, $\hat{\rho}_{N+1}^V = \hat{\rho}_{N+1}^L = 0$ and $\hat{\rho}_i^R = 0$, $i = 1,\ldots,N$. Thus the system to be modeled admits mass transfer between fluid phases but excludes such phenomena as adsorption, rock dissolution, and intraphase chemical reactions.

In porous media, fluid phase velocities are typically more accessible to measurement than constituent velocities, and therefore it is common to rewrite the balance equations (2-4) in terms of mean phase velocities $\underset{\sim}{v}^\alpha = \sum_i \rho_i^\alpha \underset{\sim}{v}_i^\alpha / \rho^\alpha$. Thus,

$$\frac{\partial}{\partial t}(\phi S_\alpha \rho^\alpha \omega_i^\alpha) + \nabla\cdot(\phi S_\alpha \rho^\alpha \omega_i^\alpha \underset{\sim}{v}^\alpha) + \nabla\cdot \underset{\sim}{j}_i^\alpha = \hat{\rho}_i^\alpha \tag{2-5}$$

for each phase α and each species i. The quantity $\underset{\sim}{j}_i^\alpha = \phi S_\alpha \rho^\alpha \omega_i^\alpha (\underset{\sim}{v}_i^\alpha - \underset{\sim}{v}^\alpha)$ is the hydrodynamic dispersion of species i with respect to the mean velocity of α.

Summing equation (2-5) over all fluid phases and imposing the constraints on ω_i^α and $\hat{\rho}_i^\alpha$ gives the species balance equations,

$$\frac{\partial}{\partial t}(\rho\omega_i) + \nabla\cdot[\phi(S_V \rho^V \omega_i^V \underset{\sim}{v}^V + S_L \rho^L \omega_i^L \underset{\sim}{v}^L)] + \nabla\cdot(\underset{\sim}{j}_i^V + \underset{\sim}{j}_i^L) = 0 \tag{2-6}$$

for $i = 1,\ldots,N$. This leaves the rock balance equation,

$$\frac{\partial}{\partial t}[(1-\phi)\rho^R] + \nabla\cdot[(1-\phi)\rho^R \underset{\sim}{v}^R] = 0$$

There now remains the issue of phase velocities. Let us assume that the response of the rock matrix to applied loads is negligible compared to the fluid motions, so that $\underset{\sim}{v}^R = 0$ and ϕ is constant. Hence only the fluid velocities merit concern. The most commonly assumed field equation for fluid velocity in a porous medium is Darcy's law, which for multiphase mixtures gives the superficial velocity of phase α as

$$\underset{\sim}{v}^\alpha = -\frac{\underset{\approx}{k} k_{r\alpha}}{\phi S_\alpha \mu^\alpha}\cdot(\nabla p_\alpha - \rho^\alpha g \nabla D),\quad \alpha = V \text{ or } L \tag{2-7}$$

Here p_α is the pressure in phase α; D is the depth relative to some datum; g is the gravitational acceleration; μ^α is the dynamic viscosity of phase α; $\underset{\approx}{k}$ is the permeability tensor of the rock matrix, and $k_{r\alpha}$ is the relative permeability of the rock to phase α under local conditions of saturation and composition. In isothermal, multicomponent, multiphase flows μ^α varies with the composition and pressure of phase α, while $k_{r\alpha}$ varies with the saturation S_α and with the interfacial tension determined by local phase compositions. Let us treat the simple case when $\underset{\approx}{k}$ reduces to a constant scalar. The rigorous development of the field equations (2-7) from the primitive momentum balance law is fairly lengthy; Prevost [25] and Bowen [6,7] review the derivation in some detail.

Although we have established the species balances, the velocity field equations, the restrictive equations (2-1) and (2-2), and constitutive relationships for $\underset{\approx}{k}$, $k_{r\alpha}$, and μ^α, there remain more variables than equations in our formulation. To correct this deficiency, we need to add several thermodynamic constraints. First, the molar density of each fluid phase obeys a constraint

(2-8) $$\rho^{\alpha} = \rho^{\alpha}(\omega_i^{\alpha},\ldots,\omega_{N-1}^{\alpha},p_{\alpha}),\ \alpha = V \text{ or } L$$

In practice this constraint takes a form of an algebraic equation of state that is implicit in ρ^{α}. Similarly the molar compositions of each fluid phase obey algebraic constraints of the form

(2-9) $$\omega_i^{\alpha} = \omega_i^{\alpha}(\omega_1,\ldots,\omega_{N-1},p_V),\ i = 1,\ldots,N-1;\ \alpha = V \text{ or } L$$

Again, in practice these constraints do not assume an explicit form but appear instead as equal-fugacity constraints of the type derived by Gibbs [13]. The saturations also must comply with equal-fugacity conditions, so that conceptually we have

(2-10) $$S_V = S_V(\omega_1,\ldots,\omega_{N-1},p_V),\ \alpha = V \text{ or } L$$

This relationship often appears as a constraint on phase mole fractions, defined by the relationship $Y_{\alpha} = \rho^{\alpha}S_{\alpha}/(\rho^V S_V + \rho^L S_L)$. Finally, the pressures in the two fluid phases must differ by an amount determined by both the local interfacial tension and the microscopic geometry of the fluid-fluid and rock-fluid phase boundaries. The capillary pressure

$$p_{cVL} = p_V - p_L$$

quantifies the net macroscopic effects of these influences.

As mentioned, the thermodynamic constraints on ρ^{α}, ω_i^{α}, and S_V assume numerical forms based on classical thermostatics. A phenomenological equation of state such as that proposed by Peng and Robinson [23] can be particularly useful in this regard, since it offers a single, consistent source of data correlating density variations and vapor-liquid phase behavior in the fluid system. For such quantities as ϕ, $\underset{\approx}{k}$, $k_{r\alpha}$, p_{cVL}, μ^{α}, and $\underset{\sim}{j}_i^{\alpha}$, on the other hand, it is necessary to use empirical correlations specific to the reservoir under study. These correlations can take the form of best-fit curves, or they can be based on interpolation between measured data. The hydrodynamic dispersions $\underset{\sim}{j}_i^{\alpha}$ pose great difficulty in this regard, since they are both difficult to measure and not well understood theoretically. Conventional wisdom seems to hold that dispersion contributes negligibly to the transport phenomena in multiphase reservoir flows, so compositional models customarily omit the dispersive fluxes $\nabla\cdot(\underset{\sim}{j}_i^V + \underset{\sim}{j}_i^L)$ (Aziz and Settari, [3], Section 12.5).

3. The Governing Equations. The balance laws, velocity field equations, restrictive equations, and thermodynamic constraints form a system of equations governing compositional flows in porous media. It is useful to assemble these relationships into a formal equation set and to review the mathematical properties of some simplified versions of this governing system.

Substituting the velocity field equations (2-7) into the species balance equations (2-6) and neglecting dispersion and gravity gives

$$(3\text{-}1)\qquad \frac{\partial}{\partial t}(\rho\omega_i) - \nabla\cdot(\underset{\approx V}{\Lambda}\rho^V\omega_i^V\nabla p_V + \underset{\approx L}{\Lambda}\rho^L\omega_i^L\nabla p_L) = 0,\ i = 1,\ldots,N$$

where $\underset{\approx\alpha}{\Lambda} = kk_{r\alpha}/\mu^\alpha$ is the mobility of fluid phase α. This flow equation is three-dimensional in the space variables, as are petroleum reservoirs in practice. Henceforth, however, let us treat the flow field as spatially one-dimensional, realizing that ultimately the numerical methods used must be extendable to more general flows. Integrating (3-1) along the directions of uniformity, denoting the cross-sectional area by A(x), and using the capillary pressure $p_{cVL} = p_V - p_L$ then yields

$$(3\text{-}2)\qquad A\frac{\partial}{\partial t}(\rho\omega_i) - \frac{\partial}{\partial x}[(T_V\omega_i^V + T_L\omega_i^L)\frac{\partial p_V}{\partial x} - T_L\omega_i^L\frac{\partial p_{cVL}}{\partial x}] = 0,\ i = 1,\ldots,N$$

where $T_\alpha = A\Lambda_\alpha\rho^\alpha$ is the transmissibility of fluid phase α.

In addition to the N flow equations (3-2) we have the four restrictive equations (2-1) and (2-2); the $2N+1$ thermodynamic constraints (2-8), (2-9), and (2-10); the definitions of ρ and ω_i, and presumably, constitutive laws sufficient to determine T_V, T_L, and p_{cVL}. Given the problem geometry (A(x) and D(x)) and appropriate boundary and initial data, the equations just cited comprise a set of $3N+5$ equations in the $3N+5$ unknowns $\{\omega_1,\ldots,\omega_N,\omega_1^V,\ldots,\omega_N^V,\omega_1^L,\ldots,\omega_N^L,p_V,S_V,S_L,\rho^V,\rho^L\}$.

It is possible for flows governed by conservation equations to exhibit discontinuities in densities or velocities, in which case the continuum-mechanical formulation must include jump conditions as well as differential balance laws (see Eringen and Ingram [11]). The possibility of discontinuous flows implies that we may not be able to demand classical solutions to the governing differential equations. Rather, we may have to settle for weak solutions. If we write the governing system in the abbreviated form $\frac{\partial f_1}{\partial x}(\omega) + \frac{\partial f_2}{\partial x}(\omega) = 0$ on an (x,t)-domain $\Omega\times\Theta$, then the criterion that ω be a weak solution is as follows: for any function $\psi\in C^\infty(\Omega\times\Theta)$ having compact support,

$$\int_{\Omega\times\Theta}[f_1(\omega)\frac{\partial\psi}{\partial t} + f_2(\omega)\frac{\partial\psi}{\partial x}]dx\,dt = 0$$

(see Birkhoff [4]).

Whether a particular governing system indeed has discontinuous solutions is in general a difficult question. It is possible, however, to show that certain natural simplifications of the governing system admit discontinuous solutions. For example, if $N = 2$ and neither species is shared between phases ($\omega_1^L = \omega_2^V = 0$, say), and if in addition the effects of capillarity together with porosity and density variations are negligible, then the flow equations (3-2) reduce to

$$\phi \frac{\partial S_V}{\partial t} + \frac{\partial q_V}{\partial x} = 0$$

$$\phi \frac{\partial}{\partial t}(1 - S_V) + \frac{\partial q_L}{\partial x} = 0$$

in a geometry where A is uniform. Here $q_\alpha = -\Lambda_\alpha \partial p_V / \partial x$ is the flow rate of fluid phase α. If the total flow rate $q = q_V + q_L$ is constant, then only one of these equations is independent, and it can be written as the single hyperbolic conservation law

$$\text{(3-3)} \qquad \frac{\partial S_V}{\partial t} + \frac{\partial}{\partial x}(q\phi^{-1} f_V) = 0$$

where $f_V = \Lambda_V / (\Lambda_V + \Lambda_L) = f_V(S_V)$ denotes the fractional flow function. Equation (3-3) is the Buckley-Leverett equation.

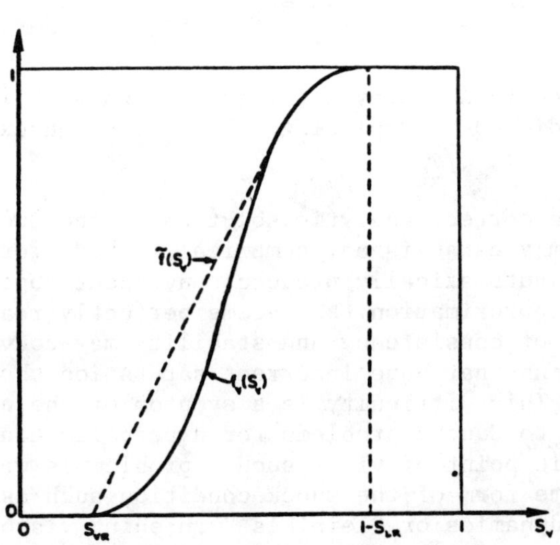

FIGURE 1 A typical fractional flow function $f_V(S_V)$ and its convex hull $\tilde{f}(S_V)$.

Equation (3-3) is one of the fundamental equations of petroleum reservoir mechanics, and it has attracted attention for decades beginning with Buckley and Leverett [8]. By now it is well known that a large class of physically realistic fractional flow functions f_V -- namely those that are nonconvex over their support -- can lead to shock-like saturation fronts that can only be weak solutions to the hyperbolic conservation law (3-3). Figure 1 shows such a nonconvex function f_V along with its convex hull $\tilde{f}$. Figure 2 shows the triple-

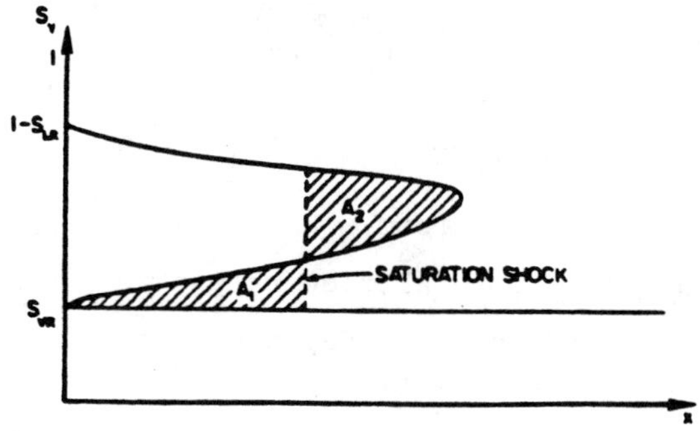

FIGURE 2 Triple-valued saturation predicted by the Buckley-Leverett characteristic equation.

valued solution S_V, gotten by interpreting the characteristic equation

$$\left.\frac{dx}{dt}\right|_{S_V} = q\phi^{-1}\frac{df_V}{dS_V}$$

literally for certain Cauchy data, and the physically correct saturation shock predicted by replacing f_V with its convex hull following Welge [28].

Although the correct analytic solution to the Buckley-Leverett problem is firmly established, numerical methods for solving this problem do not automatically produce convergent approximations. In fact, a naive approximation that seems perfectly reasonable by the usual criteria of consistency and stability may converge to mass-conserving solutions that have incorrect saturation shocks (Allen and Pinder, [2]). This difficulty is a symptom of the nonuniqueness of weak solutions to Cauchy problems for hyperbolic conservation laws. From an analytic point of view, such a problem is well posed only if it includes some form of the shock condition such as the entropy condition of gas dynamics or Oleinik's "vanishing viscosity" constraint [21]. The Welge tangent construction [28] is a geometric equivalent of this condition. A numerical method for solving this type of Cauchy problem will converge only if it imposes a discrete analog of the shock condition.

In recent years several investigators have extended the Buckley-Leverett theory to porous-media flows of greater complexity. Isaacson [15] and Temple [27], for example, study a pair of conservation laws governing simple polymer floods, in which two-phase flow accompanies the passive transport of a viscosity-altering solute. Helfferich [14] presents a generalization of the Buckley-Leverett theory to multicomponent flows with interphase mass transfer. Under a set of assumptions analogous to those underlying equation (3-3), Helfferich develops the following conservation equations for a system of N species:

$$\frac{\partial \omega_i}{\partial t} + \frac{\partial}{\partial x}(q\phi^{-1} f_i) = 0, \quad i = 1,\ldots,N-1$$

where $f_i = f_V \omega_i^V + f_L \omega_i^L$. Helfferich then shows that this system of hyperbolic conservation laws admits weak solutions which can exhibit up to $N-1$ independent but coherent shocks for physically reasonable choices of the flux functions f_i.

These observations have implications for the numerical solution of the full system of governing equations. While in reality models of compositional flows must account for capillarity, the contribution of capillary pressure gradients to the flow field is often small compared with convection driven by applied pressure gradients. In such cases the governing flow equations are effectively singular perturbations of a system of hyperbolic equations, and the dissipative effect of capillarity smears shocks into small regions of steep composition gradients. While there may be no problem with uniqueness in the true solutions to these equations, the effects of capillarity may be too small to guarantee practical convergence of certain discrete approximations to physically correct solutions. Therefore any numerical method for solving the flow equations (3-2) should impose some form of the shock condition. In finite-difference models the use of upstream-weighted difference approximations to convective fluxes is nearly universal (see, for example, Coats [9] and Nghiem et al. [19]) since it induces truncation errors that act like vanishing viscosities or, for these physics, "vanishing capillarities." In the next section we shall examine an analogous technique for use with finite-element collocation.

4. A Collocation Method. Having established the system of equations governing compositional flows, let us turn to the task of solving these equations numerically. The primary purpose of this section is to discuss a method of collocation on finite elements that is suited to modeling the convection-dominated flows that occur in oil reservoirs under miscible gas flooding. After discussing the method in a fairly general setting, we shall briefly examine the collocation solutions to some simple analogs of the governing flow equations. Section 5 presents a specific scheme for the fully compositional case.

Consider a partial differential equation of the form

$$\frac{\partial u}{\partial t} + Eu = 0 \tag{4-1}$$

defined on a space-time domain $\Omega \times T = [0, x_{max}] \times [0, t_{max}]$. Here, for purposes of illustration, let E be a spatial operator of the form

$$E = -\frac{\partial}{\partial x}[\alpha_1(x)\frac{\partial}{\partial x}] + \alpha_2(x)\frac{\partial}{\partial x}$$

where $\alpha_1, \alpha_2 \in C^\infty(\Omega)$. Thus (4-1) is an idealization of the general flow equation (3-2).

When discussing finite element methods it is appropriate to view (4-1) in its variational form. To do this, we allow $u(\cdot,t)$ to lie in the Sobolev space $H^1(\Omega)$ containing real-valued functions f on Ω such that both $f, df/dx \in L^2(\Omega)$. Such a function is a solution to the variational problem associated with (4-1) provided (1) $u(\cdot,t) \in H^1(\Omega)$ for all $t \in \Theta$; (2) $u(x,\cdot) \in C^1(\Theta)$ for all $x \in \Omega$; (3) $u(x,t)$ satisfies the auxiliary data, and (4)

$$\int_\Omega \frac{\partial u}{\partial t}\psi\, dx + \int_\Omega \left(\alpha_1 \frac{\partial u}{\partial x} + \alpha_2 u\right)\frac{\partial \psi}{\partial x}\, dx = 0 \tag{4-2}$$

for any test function $\psi \in H^1(\Omega)$ that vanishes on the boundary $\partial\Omega$.

To construct a discrete approximation to the spatial part of the problem (4-2), let Δ_M: $0 = \overline{x}_1 < \cdots < \overline{x}_M = x_{max}$ be a partition of Ω. For simplicity we shall assume that Δ_M is uniform with mesh $\Delta x = \overline{x}_\ell - \overline{x}_{\ell-1}$, although this is neither necessary nor always desirable in applications. Associated with the partition Δ_M is a finite-dimensional subspace of $H^1(\Omega)$, namely,

$$H_3(\Delta_M) = \{\gamma \in H^1(\Omega) \mid \gamma \text{ is a cubic polynomial over each interval } [\overline{x}_\ell, \overline{x}_{\ell+1}]\}$$

This subspace $H_3(\Delta_M)$ is the span of the Hermite cubic interpolation basis $\{H_{0,\ell}(x), H_{1,\ell}(x)\}_{\ell=1}^M$ whose elements are the piecewise polynomials depicted in Figure 3 (see Prenter [24], Chapter 3). Thus every $\gamma \in H_3(\Delta_M)$ is a finite linear combination

$$\gamma(x) = \sum_{\ell=1}^{M} [\gamma_\ell H_{0,\ell}(x) + \gamma'_\ell H_{1,\ell}(x)]$$

where the unique coefficients γ_ℓ, γ'_ℓ have the values $\gamma(\overline{x}_\ell)$, $d\gamma(\overline{x}_\ell)/dx$, respectively. As $\Delta x \to 0$, $H_3(\Delta_M)$ is dense as a subspace of $H^1(\Omega)$ in the Sobolev norm $\|\cdot\|_{2,1}$, so that finite-element approximations in $H_3(\Delta_M)$

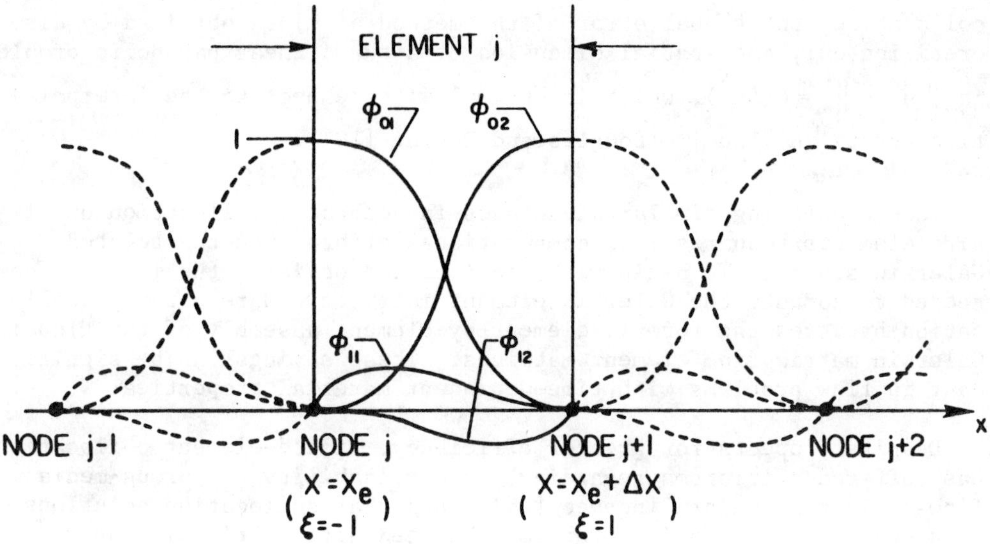

FIGURE 3 Hermite basis functions for several adjacent elements.

are consistent (Oden and Reddy [20], Section 8.3). Finally, there is an interpolation operator associated with $H_3(\Delta_M)$, namely the projection I_M: $C^1(\Omega) \to H_3(\Delta_M)$ defined by

$$(I_M f)(x) = \sum_{\ell=1}^{M} [f(\overline{x}_\ell)H_{0,\ell}(x) + \frac{df}{dx}(\overline{x}_\ell)H_{1,\ell}(x)]$$

In terms of this formalism, the spatially discrete analog to equations (4-1) and (4-2) is

$$(4\text{-}3) \qquad \int_\Omega \left[\frac{\partial \hat{u}}{\partial t}(x,t) + E\hat{u}(x,t)\right] \delta_k(x)\, dx = 0, \quad k = 1,\ldots,2M-2$$

where $\hat{u}(\cdot,t) \in H_3(\Delta_M)$ is unknown and $\{\delta_k\}_{k=1}^{2M-2}$ is a collection of weight functions. Assuming we have an appropriate set of boundary data for $\hat{u}$, equations (4-3) furnish a set of ordinary differential equations governing the evolution of the coefficients u_ℓ, u'_ℓ in time. The initial data for this system are, in the simplest case, just projections of the exact initial data onto $H_3(\Delta_M)$: $\hat{u}(x,0) = I_M u_0(x)$, $x \in (0,x_{max})$.

Collocation means choosing as weight functions the Dirac distributions $\delta_k(x) = \delta(x - x_k)$, where the $2M-2$ collocation points x_k are spread throughout Ω. de Boor and Swartz [5] and Douglas and Dupont [10] show that the highest order of accuracy in Δx for this method results when the collocation points x_k are the Gauss points $\overline{x}_\ell + \frac{\Delta x}{2}(1 \pm 1/\sqrt{3})$, $\ell = 1,\ldots,M-1$. In this case, called orthogonal collocation, the global error of the method of lines obtained by discretizing only the spatial dimension of a quasilinear parabolic problem is $\| u - \hat{u} \|_\infty = O(\Delta x^4)$, which is optimal with respect to the interpolation error $\| u - I_M u \|_\infty$ (Douglas and Dupont [10]).

Though offering similar advantages in accuracy, collocation on finite elements requires less computational effort than the related Galerkin scheme. To begin with, collocation obviates the integrations needed to compute the Galerkin element matrices. More notably, collocation bypasses the formal, element-by-element assembly of the global Galerkin matrix from element matrices. These savings can be significant in flow problems with time-dependent material properties.

Despite propects for greater efficiency, finite-element collocation has suffered criticisms regarding its applicability to porous-media flows. In particular Sincovec [26] found that collocation solutions to a convection-diffusion problem exhibited difficulties in the presence of steep concentration gradients and that his collocation schemes failed to yield convergent approximations to the hyperbolic Buckley-Leverett problem.

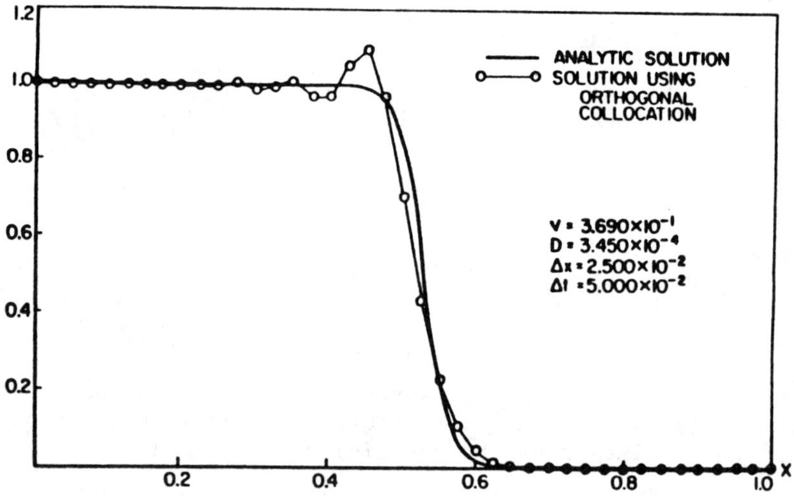

FIGURE 4 Solution to the convection-dispersion equation using orthogonal collocation.

These criticisms are, in a sense, justified. Figure 4, for example, shows a numerical solution profile to a convection-diffusion problem

$$\frac{\partial \omega}{\partial t} + v \frac{\partial \omega}{\partial x} - D \frac{\partial^2 \omega}{\partial x^2} = 0 \qquad (4\text{-}4)$$

$$\omega(x,0) = 0, \ x \in (0,1)$$

$$\omega(0,t) = 1, \ t \geq 0$$

$$\frac{\partial \omega}{\partial x}(x_{max},t) = 0, \ t \geq 0$$

using orthogonal collocation in space and a Crank-Nicolson difference scheme in time. This solution exhibits spurious oscillations upstream of the sharp front in the exact solution, thereby violating the maximum principle of parabolic equations (John [17], Section 7.1). These oscillations are typical of high-order spatial approximations to convection-dominated problems where the grid Peclet number $Pe_{\Delta x} = v\Delta x/D$ exceeds a critical value of order unity (Jensen and Finlayson [16]).

While oscillatory solutions to the convection-dispersion equation may be problematic, orthogonal collocation can give solutions to the Buckley-Leverett problem that are undeniably wrong. Consider the nonlinear Cauchy problem

$$\frac{\partial S_V}{\partial t} + \frac{\partial}{\partial x}[q\phi^{-1} f_V(S_V)] \text{ on } [0,\infty) \times [0,\infty) \tag{4-5}$$

$$S_V(0,t) = 1 - S_{LR}, \ t \geq 0$$

$$S_V(x,0) = S_{VR}, \ x > 0$$

where $f_V(S_V)$ is the fractional flow function shown in Figure 1. Figure 5 shows a numerical solution profile to this problem using a linearized implicit orthogonal collocation scheme (see Allen and Pinder

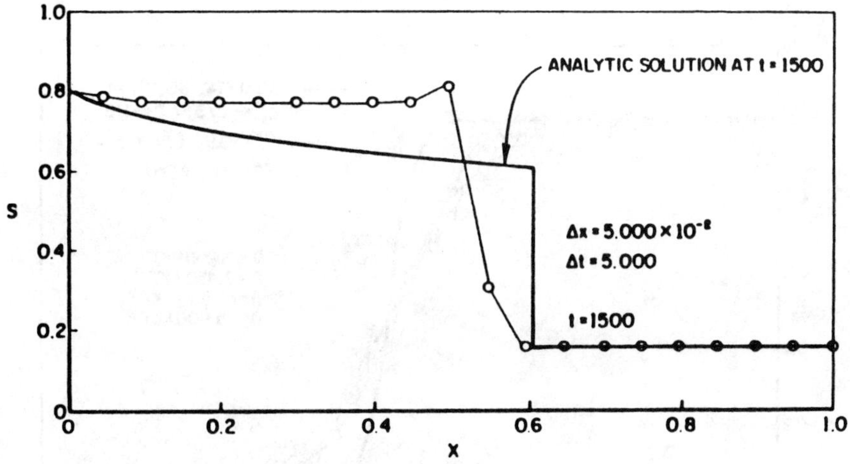

FIGURE 5 Solution to the Buckley-Leverett problem using orthogonal collocation.

[2] for details). The figure also shows the true solution. The

numerical solution is qualitatively flawed in that it predicts a flood front that is too strong and too late, even though the scheme conserves mass. In contrast to the spurious wiggles arising in the convection-dispersion equation, the incorrect shock predicted in the Buckley-Leverett problem is unacceptable for any design purposes.

A technique called upstream collocation (Allen and Pinder [2], Allen [1]) offers one way around these problems. The basic idea behind this method is to split the spatial operator E into a "convective" (first order) flux and a non-convective flux, or $E = C + B$, and then to evaluate the convective flux C at points upstream of the Gauss points. Thus, for example, with the convection-diffusion equation (4-4), $C = v\partial\omega/\partial x$ and $B = -D\partial^2\omega/\partial x^2$, so the collocation equations take the form

$$\text{(4-6)} \quad \frac{\partial\hat{\omega}}{\partial t}(x_k,t) + v\,\frac{\partial\hat{\omega}}{\partial x}(x_k^*,t) - D\,\frac{\partial^2\hat{\omega}}{\partial x^2}(x_k,t) = 0, \quad k = 1,\ldots,2M-2$$

Here $\hat{\omega}$ represents the unknown Hermite cubic trial function and x_k^* signifies a point $x_k - \zeta\Delta x$ located upstream of the Gauss point x_k but lying in the same element $[\bar{x}_\ell, \bar{x}_{\ell+1}]$. It is easy to show that this procedure induces an error that augments the diffusion term by an amount proportional to Δx. In fact, (4-6) is equivalent to

$$\frac{\partial\hat{\omega}}{\partial t}(x_k,t) + v\,\frac{\partial\hat{\omega}}{\partial x}(x_k,t) - (D + \zeta v\Delta x)\frac{\partial^2\omega}{\partial x^2}(x_k,t) = O(\Delta x^2), \quad k = 1,\ldots,2M-2$$

This numerically augmented diffusion shows up as smearing in the approximate solution to the convection-diffusion problem (4-4) as Figure 6 demonstrates. The parameters ξ_1^*, ξ_2^* in this figure are loci of the

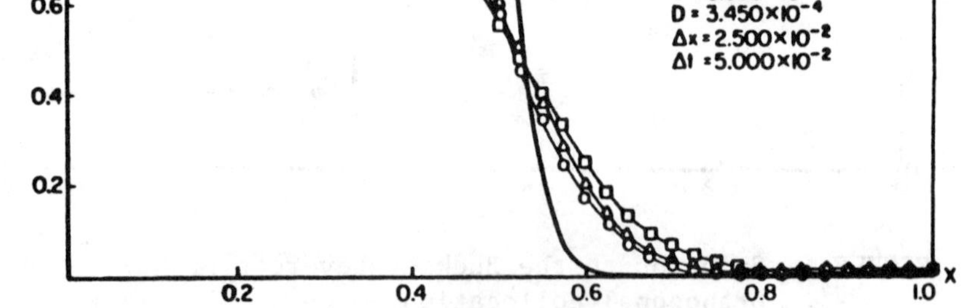

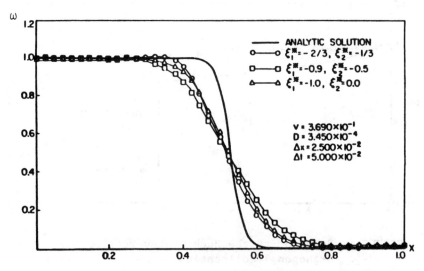

FIGURE 6 Solutions to the convection-dispersion equation using several choices of upstream collocation points.

upstream collocation points in terms of the local element coordinate $\xi(x) = [2(x - \overline{x}_\ell)/\Delta x] - 1$.

With the Buckley-Leverett problem (4-5) the effects of upstream collocation are more dramatic. Here $C = \partial(q\phi^{-1}f_V)/\partial x$ and $B = 0$, so the collocation equation is

$$\frac{\partial \hat{S}}{\partial t}(x_k, t) + \frac{\partial}{\partial x}[q\phi^{-1}\hat{f}(x_k^*, t)] = 0, \quad k = 1, \ldots, 2M-2$$

where, again, $x_k^* = x_k - \zeta\Delta x$ signifies an upstream collocation point. In this case an error analysis shows that upstream collocation induces an $O(\Delta x)$ error resembling a capillary pressure gradient in form (Allen [1]). This error is precisely the numerical analog of "vanishing capillarity," and as Figure 7 illustrates, it forces convergence to the physically correct solution of the Cauchy problem.

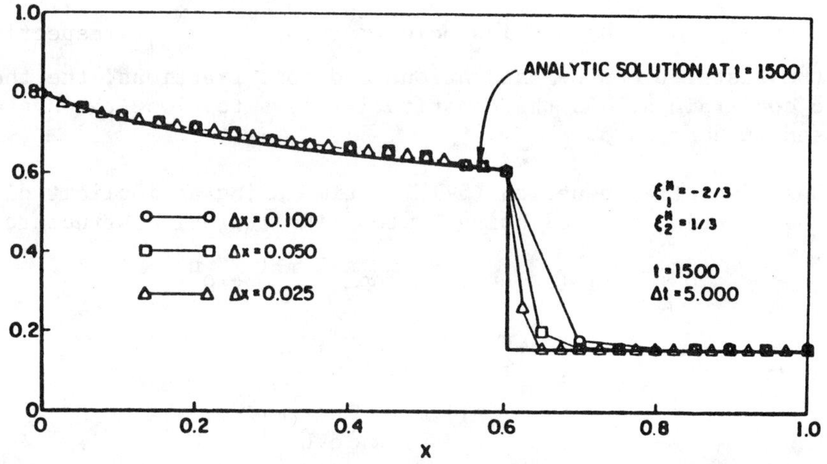

FIGURE 7 Solutions to the Buckley-Leverett problem using upstream collocation with spatial grids of varying mesh.

These two simple problems argue the need for upstream weighting in more complicated convection-dominated flows. While the artificial dissipation resulting from upstream-biased approximations may be worrisome, the failure to produce convergent numerical schemes in problems with sharp material interfaces is a more drastic shortcoming. Let us turn now to the implementation of upstream collocation in a simple compositional model.

5. <u>A Compositional Model</u>. Consider the simultaneous flow of N molecular species distributed between two fluid phases V and L in a porous medium, as described in Sections 2 and 3. An implicit pressure-

explicit composition scheme for simulating such flows requires the derivation of an equation governing the pressure in one of the fluids, say V, and a scheme for coupling an implicit solution technique for the pressure equation with the remaining N - 1 independent species transport equations, solved explicitly at each iteration.

The pressure equation results from summing the species transport equations (3-2) over all N components and using the restrictions (2-1) and (2-2):

$$A\frac{\partial\rho}{\partial t} - \frac{\partial}{\partial x}\left(T_T\frac{\partial p_V}{\partial x} - T_L\frac{\partial p_{cVL}}{\partial x}\right) = 0 \tag{5-1}$$

where $T_T = T_V + T_L$. We then have N - 1 independent species transport equations:

$$A\frac{\partial}{\partial t}(\rho\omega_i) - \frac{\partial}{\partial x}\left(T_i\frac{\partial p_V}{\partial x} - T_L\omega_i^L\frac{\partial p_{cVL}}{\partial x}\right) = 0,\ i = 1,\ldots,N-1 \tag{5-2}$$

where $T_i = T_V\omega_i^V + T_L\omega_i^L$. We can regard (5-1) and (5-2) as equations for the pressure p_V and the species mole fractions $\{\omega_i\}_{i=1}^{N-1}$, respectively, using the restrictions on saturations and mole fractions, the thermodynamic constraints, and the constitutive laws to close the system as discussed in Section 3.

Let us discretize equation (5-1) in time using an implicit difference scheme having the following Newton-like iterative structure:

$$\begin{aligned}\rho^{n+1,m} &+ (\Delta\rho/\Delta p_V)^{n+1,m}\delta p_V^{n+1,m+1} - \rho^n \\ &= \tau\frac{\partial}{\partial x}\left[T_T^{n+1,m}\frac{\partial}{\partial x}\left(p_V^{n+1,m} + \delta p_V^{n+1,m+1}\right)\right. \\ &\qquad\left. - T_L^{n+1,m}\frac{\partial}{\partial x}p_{cVL}^{n+1,m}\right]\end{aligned} \tag{5-3}$$

where $\tau = \Delta t/A$. In this equation, $(\cdot)^n$ stands for the value of $(\cdot)$ at the known time level $n\Delta t$ after numerical convergence of the iterations, and $(\cdot)^{n+1,m}$ stands for the most recently computed iterate of $(\cdot)$ at the unknown time level $(n+1)\Delta t$. The quantity $\delta p_V^{n+1,m+1}$ is the $(m+1)$-st iterative correction to the pressure at time $(n+1)\Delta t$, so $p_V^{n+1,m+1} = p_V^{n+1,m} + \delta p_V^{n+1,m+1}$. The factor $(\Delta\rho/\Delta p_V)$ is an approximation to $d\rho/dp_V$ computed as follows:

$$\frac{\Delta\rho}{\Delta p_V} = \frac{1}{\Delta p_V}[\rho(\omega_1,\ldots,\omega_{N-1},p_V + \Delta p_V) - \rho(\omega_1,\ldots,\omega_{N-1},p_V)]$$

where, in practice, $\Delta p_V \simeq 10^2$ Pa.

To discretize the composition equations (5-2) in time, let us use the following Euler-like scheme:

$$\Delta_t \omega_i^{n+1,m+1} = \left\{ \tau \frac{\partial}{\partial x}\left[T_i^{n+1,m} \frac{\partial}{\partial x} p_V^{n+1,m+1} - (T_L \omega_i^L)^{n+1,m} \frac{\partial}{\partial x} p_{cVL}^{n+1,m} \right] - \omega_i^n \Delta_t \rho^{n+1,m+1} \right\} / \rho^{n+1,m+1} \tag{5-4}$$

where $\Delta_t(\cdot)^{n+1,m+1} = (\cdot)^{n+1,m+1} - (\cdot)^n$. The updated values of total fluid density appearing here are available from the latest iteration of equation (5-3):

$$\rho^{n+1,m+1} = \tau \frac{\partial}{\partial x}\left(T_T^{n+1,m} \frac{\partial}{\partial x} p_V^{n+1,m+1} - T_L^{n+1,m} \frac{\partial}{\partial x} p_{cVL}^{n+1,m} \right) + \rho^n$$

There remains the task of discretizing (5-3) and (5-4) in space. Following the discussion in Section 4, let us consider the projections of unknown pressures and compositions onto trial spaces of Hermite cubics:

$$\delta p_V \simeq \delta \hat{p} = \sum_{\ell=1}^{M} (\pi_\ell H_{0,\ell} + \pi'_\ell H_{1,\ell})$$

$$\Delta_t \omega_i \simeq \Delta_t \hat{\omega}_i = \sum_{\ell=1}^{M} (w_{i,\ell} H_{0,\ell} + w'_{i,\ell} H_{1,\ell})$$

$$p_V \simeq \hat{p} = \sum_{\ell=1}^{M} (\Pi_\ell H_{0,\ell} + \Pi'_\ell H_{1,\ell})$$

$$\omega_i \simeq \hat{\omega}_i = \sum_{\ell=1}^{M} (W_{i,\ell} H_{0,\ell} + W'_{i,\ell} H_{1,\ell})$$

For the dependent variables ρ, $\Delta\rho/\Delta p_V$, T_T, T_L, T_i, and $(T_L \omega_i^L)$, let us project to spaces of C^0 piecewise linear Lagrange interpolates using the nodal values. For example, the approximation for the total fluid density becomes

$$\rho(\omega_1(x),\ldots,\omega_{N-1}(x),p_V(x)) \simeq \sum_{\ell=1}^{M} \rho(\overline{x}_\ell) L_\ell(x)$$

where $\rho(\overline{x}_\ell)$ denotes the nodal value $\rho(W_{1,\ell},\ldots,W_{N-1,\ell},\Pi_\ell)$ and $\{L_\ell(x)\}_{\ell=1}^{M}$ is the basis for piecewise linear interpolation between nodes. For the capillary pressure p_{cVL}, the implied presence of $\partial^2 p_{cVL}/\partial x^2$ in the flow equations makes a piecewise quadratic Lagrange representation more appropriate: on each element $[\overline{x}_\ell, \overline{x}_{\ell+1}]$, let

$$\hat{p}_{cVL}(x) = p_{cVL}(\overline{x}_\ell) Q_{\ell,1}(x) + p_{cVL}(\overline{x}_{\ell+\frac{1}{2}}) Q_{\ell,2}(x) + p_{cVL}(\overline{x}_{\ell+1}) Q_{\ell,3}(x)$$

where $\overline{x}_{\ell+\frac{1}{2}} = (\overline{x}_\ell + \overline{x}_{\ell+1})/2$ and $\{Q_{\ell,1}, Q_{\ell,2}, Q_{\ell,3}\}$ is the Lagrange quadratic interpolating basis for the element containing nodes $\overline{x}_\ell$, $\overline{x}_{\ell+\frac{1}{2}}, \overline{x}_{\ell+1}$.

Substituting these finite-element spatial representations into the time-differenced equations (5-3) and (5-4) and collocating gives a set of linear algebraic equations at each iterative level. For the pressure equation (5-1), these are

$$\widehat{\left(\frac{\Delta\rho}{\Delta p_V}\right)}^{n+1,m}(x_k)\,\delta\hat{p}^{n+1,m+1}(x_k) - \tau\left[\frac{\partial}{\partial x}\hat{T}_T^{n+1,m}(x_k^*)\frac{\partial}{\partial x}\delta\hat{p}^{n+1,m+1}(x_k^*) + \hat{T}_T^{n+1,m}(x_k)\frac{\partial^2}{\partial x^2}\delta\hat{p}^{n+1,m+1}(x_k)\right]$$

$$= -\Big\{\hat{\rho}^{n+1,m}(x_k) - \hat{\rho}^n(x_k) - \tau\Big[\frac{\partial}{\partial x}\hat{T}_T^{n+1,m}(x_k^*)\frac{\partial}{\partial x}\hat{p}^{n+1,m}(x_k^*) + \hat{T}_T^{n+1,m}(x_k)\frac{\partial^2}{\partial x^2}\hat{p}^{n+1,m}(x_k) - \frac{\partial}{\partial x}\hat{T}_L^{n+1,m}(x_k)\frac{\partial}{\partial x}\hat{p}_{cVL}^{n+1,m}(x_k) - \hat{T}_L^{n+1,m}(x_k)\frac{\partial^2}{\partial x^2}\hat{p}_{cVL}^{n+1,m}(x_k)\Big]\Big\}, \tag{5-5}$$

$$k = 1,\ldots,2M-2$$

where, as in Section 4, x_k^* denotes points upstream of the usual collocation points x_k in each element. The notation "^" here signifies projection onto the appropriate finite-element subspace. The discrete analogs of the composition equations (5-2) are

$$\Delta_t\hat{\omega}_i^{n+1,m+1} = \Big\{\tau\Big[\frac{\partial}{\partial x}\hat{T}_i^{n+1,m}(x_k^*)\frac{\partial}{\partial x}\hat{p}^{n+1,m+1}(x_k^*) + \hat{T}_i^{n+1,m}(x_k)\frac{\partial^2}{\partial x^2}\hat{p}^{n+1,m+1}(x_k) - \frac{\partial}{\partial x}\widehat{(T_L\omega_i^L)}^{n+1,m}(x_k)\frac{\partial}{\partial x}\hat{p}_{cVL}^{n+1,m} - \widehat{(T_L\omega_i^L)}^{n+1,m}(x_k)\frac{\partial^2}{\partial x^2}\hat{p}_{cVL}^{n+1,m}\Big] - \hat{\omega}_i^n(x_k)\Delta_t\hat{\rho}^{n+1,m+1}(x_k)\Big\}\cdot\frac{1}{\hat{\rho}^{n+1,m+1}(x_k)}, \tag{5-6}$$

$$i = 1,\ldots,N-1;\; k = 1,\ldots,2M-2$$

Given appropriate auxiliary conditions, these two sets of discrete equations are equivalent to N matrix equations of order (2M - 2) at each iterative level. The pressure equations (5-5) reduce to

$$[M]^{n+1,m}\{\pi\}^{n+1,m+1} = -\{R\}^{n+1,m}$$

where $\{\pi\}^{n+1,m+1}$ denotes a column vector containing the Hermite coefficients $\{\pi_\ell^{n+1,m+1}, \pi_\ell'^{n+1,m+1}\}_{\ell=1}^M$, and the composition equations (5-6) take the form

$$[B]\{w_i\}^{n+1,m+1} = \{r_i\}^{n+1,m}, \quad i = 1,\ldots,N-1$$

where $\{w_i\}^{n+1,m}$ is the column vector of coefficients $\{\omega_{i,\ell}^{n+1,m+1}, \omega_{i,\ell}'^{n+1,m+1}\}_{\ell=1}^M$ and [B] is the matrix of Hermite cubic interpolation coefficients $H_{j,\ell}(x_k)$, $j=0,1$, associated with the Gauss points. Since [B] is constant it is necessary to invert it only once, during the initialization segment of a simulator.

The flow equations and their algebraic analogs require N initial conditions, specified numerically as follows:

$$\hat{p}(x,0) = p_0(x), \quad x \in (0,x_{max})$$

$$\hat{\omega}_i(x,0) = \omega_{i,0}(x), \quad x \in (0,x_{max}), \quad i = 1,\ldots,N-1$$

where the right sides of these equations lie in the trial space $H_3(\Delta_M)$. For boundary data, let us specify the injection rate $q_I(t)$ and injected fluid composition at $x=0$ and a specified producing pressure and zero dispersive flux at $x=x_{max}$:

$$\frac{\partial\hat{p}}{\partial x}(0,t) = -q_I(t)/T_T(0,t), \quad t \geq 0$$

$$\hat{\omega}_i(0,t) = \omega_I(t), \quad t \geq 0, \quad i = 1,\ldots,N-1$$

$$\hat{p}(x_{max},t) = p_r(t), \quad t \geq 0$$

$$\frac{\partial\hat{\omega}_i}{\partial x}(x_{max},t) = 0, \quad t \geq 0, \quad i = 1,\ldots,N-1$$

The flowchart drawn in Figure 8 illustrates the general structure of the model. The segment labeled "UPDATE COEFFICIENTS..." deserves some comment. This is the portion of the code that computes, among other things, the instantaneous thermodynamics of the fluid system at each node $\overline{x}_\ell$. As mentioned in Section 2, the thermodynamic constraints assume the form of an equation of state together with nonlinear algebraic equations forcing equality between the local fugacities in the vapor and liquid. The conventional approach to these constraints is to solve the equal-fugacity equations during simulation time using some iterative scheme (see Coats [9], Nghiem et al. [19]). This

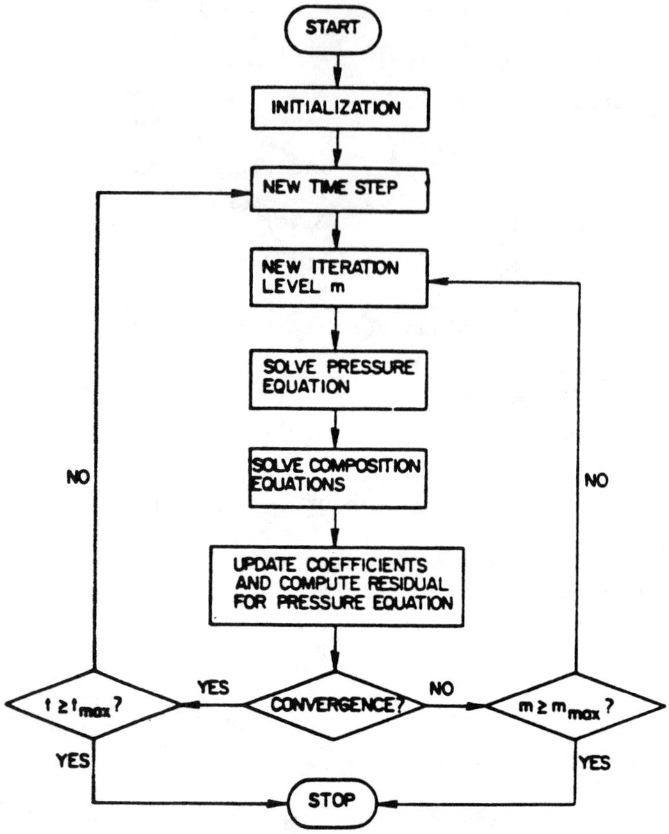

FIGURE 8 Flowchart of the overall structure of the compositional simulator.

approach becomes very costly as the number N of molecular species grows, and, what is more, the calculations often exhibit poor convergence near critical points of the fluid mixture.

One relatively simple way to circumvent these difficulties is to construct interpolatory approximations of the manifolds (such as saturation pressure domes) separating different phase regimes and of the tie lines defining phase equilibria, in advance of running the simulator. If the data for the interpolations are based on an equation of state and one takes pains to preserve thermodynamic admissibility near critical loci, then it is possible to maintain thermodynamic consistency while incurring errors that are small compared with those inherent in the equation of state. Furthermore, the use of such an interpolation scheme allows order-of-magnitude or greater reductions in the computational effort spent on thermodynamics and can eliminate convergence difficulties during simulation. This approach should be broadly applicable; for a particular implementation see Allen [1].

Let us consider a sample problem for the ternary system $CO_2 + n\text{-}C_4H_{10} + n\text{-}C_{10}H_{22}$ at 344.36 K. While this mixture is considerably simpler than the fluid systems found in actual petroleum reservoirs, it exhibits many of the qualitative features associated with the phase behavior of CO_2 floods in fields undergoing enhanced recovery (Metcalfe and Yarborough [18], Orr and Jensen [22]). In this problem the injected fluid is a mixture lying in the vapor region of the ternary phase diagram, and the initial resident fluid is a two-phase mixture at residual vapor saturation. The injected fluid and initial reservoir fluids have compositions that are appropriate for the establishment of a vaporizing gas drive. The injection rate is 2.0 mol/s into a reservoir 12 m long, 1 m^2 in cross-section, and having permeability 1.0×10^{-12} m^2 ($\simeq$ 1 darcy) and porosity 0.2. The pressure at the producing end of the reservoir remains constant at 10 MPa. More details regarding parameter values, as well as other sample problems, appear in Allen [1].

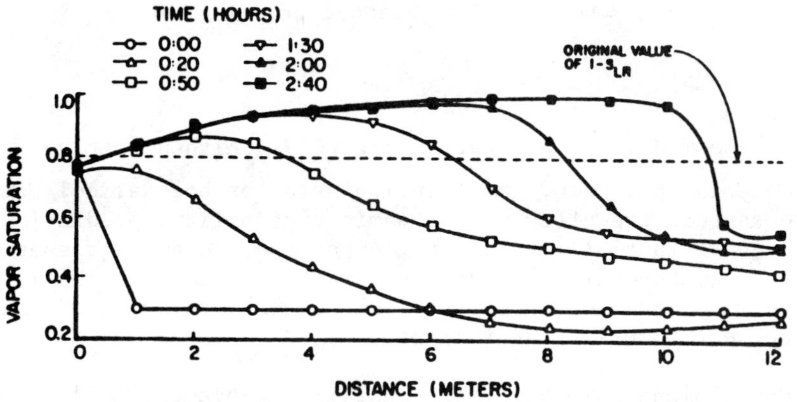

FIGURE 9 Saturation profiles for the vaporizing gas drive.

Figure 9 shows the profile of vapor saturation S_V versus distance x for this example at various times during the flood, computed using upstream collocation with $\zeta = 0.3$. As the figure displays, a transition zone between injected fluid and initial reservoir fluid moves down the pressure gradient, establishing a "miscible" zone in which the interfacial tension between phases in contact is small. As is typical of a successful vaporizing gas drive, the low-tension zone displaces the liquid very efficiently, leaving high vapor saturations in the swept zone upstream.

It is interesting to compare this solution, using upstream collocation, with the numerical solution produced by orthogonal collocation. Figure 10 shows the saturation profiles corresponding to those in Figure 9, the only difference between the two solutions being the choice

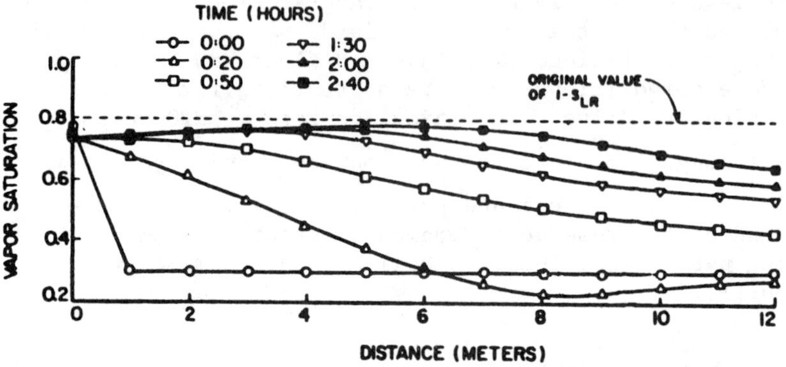

FIGURE 10 Saturation profiles for the vaporizing gas flood, as predicted by orthogonal collocation.

$x_k^* = x_k$ in the model run generating Figure 9. Although it has good global material balances, the solution using orthogonal collocation fails to capture the effects of dynamic miscibility, predicting no decrease in residual liquid saturation as the flood progresses. As in the Buckley-Leverett problem, the artificial dissipation induced by upstream collocation appears to be necessary to produce qualitatively correct numerical approximations.

6. Conclusions. We have seen that finite-element collocation can be effective in capturing the behavior of multiphase, multicomponent fluid flows in porous media. Central to the success of the method is the use of an upstream biasing technique that forces convergence to physically correct solutions through a numerical version of the vanishing viscosity condition. This upstreaming method is analogous to the upstream weighting techniques widely used in finite-difference models of convection-dominated flows.

Upstream collocation exacts a price, however: the artificial dissipation that ensures convergence shows up as numerical smearing in the approximate solutions. This "numerical diffusion" is a significant problem in models of physical processes that exhibit sharp gradients in fluid density, saturation, or composition. Much recent effort has focussed on the use of local grid refinement around steep fronts to reduce smearing while preserving convergence (see, for example,

Ewing [12]). The extension of this class of methods to finite-element collocation should help alleviate the smearing evident in Figures 6, 7, and 9.

Finally, the applications documented here are all restricted to one space dimension. The extension of upstream collocation to several space dimensions is crucial to the widespread usefulness of the method. This work is in progress.

Acknowledgements. I am greatly indebted to Professors George F. Pinder and William G. Gray of Princeton University for their guidance during my doctoral studies, which led to this work.

REFERENCES

[1] M.B. ALLEN, Collocation Techniques for Modeling Compositional Flows in Oil Reservoirs, Lecture Notes in Engineering, Springer-Verlag, Berlin, 1984.

[2] M.B. ALLEN and G.F. PINDER, Collocation simulation of multiphase porous-medium flow, Soc. Pet. Eng. J.(1983), pp. 135-142.

[3] K. AZIZ and A. SETTARI, Petroleum Reservoir Simulation, Applied Science, London, 1979.

[4] G. BIRKHOFF, Numerical fluid dynamics, SIAM Review 25(1983), pp. 1-34.

[5] R.D. DEBOOR and B. SWARTZ, Collocation at Gaussian points, SIAM Jour. Numer. Anal. 10(1973), pp. 582-606.

[6] R.M. BOWEN, Incompressible porous media models by use of the theory of mixtures, Int. J. Engrg. Sci. 18(1980), pp. 787-800.

[7] R.M. BOWEN, Compressible porous media models by use of the theory of mixtures, Int. J. Engrg. Sci. 20(1982), pp. 697-735.

[8] S.E. BUCKLEY and M.C. LEVERETT, Mechanism of fluid displacement in sands, Trans. AIME 146(1942), pp. 107-116.

[9] K. COATS, An equation-of-state compositional model, Soc. Pet. Eng. J.(1980), pp. 363-376.

[10] J. DOUGLAS and T. DUPONT, A finite-element collocation method for quasilinear parabolic equations, Math. Comp. 27(1973), pp. 17-28.

[11] A.C. ERINGEN and J.D. INGRAM, A continuum theory of chemically reacting media -- I, Int. J. Engrg. Sci. 3(1965), pp. 197-212.

[12] R.E. EWING, Problems arising in the modeling of processes for hydrocarbon recovery, Vol. 1, Research Frontiers in Applied Mathematics, SIAM, Philadelphia, 1984, pp. 3-34.

[13] J.W. GIBBS, On the equilibrium of heterogeneous substances, Trans. Conn. Acad. 3(1876), pp. 108-248 and 3(1878), pp. 343-524.

[14] F.G. HELFFERICH, Generalized Welge construction for two-phase flow in porous media in a system with limited miscibility, SPE 9730, presented at the 57th Annual Fall Technical Conference and Exhibition of the Society of Petroleum Engineers of AIME, New Orleans, September 26-29, 1982.

[15] E.L. ISAACSON, Global solution of a Riemann problem for a non-strictly hyperbolic system of conservation laws arising in enhanced oil recovery, Jour. Comp. Phys., to appear.

[16] O.K. JENSEN and B.A. FINLAYSON, Oscillation limits for weighted residual methods applied to convective diffusion equations, Int. J. Num. Meth. Engrg. 15(1980), pp. 1681-1689.

[17] F. JOHN, Partial Differential Equations, 4th ed., Springer-Verlag, New York, 1981.

[18] R.S. METCALFE and L. YARBOROUGH, The effect of phase equilibria on the CO_2 displacement mechanism, Soc. Pet. Eng. J.(1979), pp. 242-252.

[19] L.X. NGHIEM, D.K. FONG, and K. AZIZ, Compositional modeling with an equation of state, Soc. Pet. Eng. J.(1981), pp. 687-678.

[20] J.T. ODEN and J.N. REDDY, An Introduction to the Mathematical Theory of Finite Elements, John Wiley and Sons, New York, 1976.

[21] O.A. OLEINIK, Construction of a generalized solution of the Cauchy problem for a quasi-linear equation of first order by the introduction of "vanishing viscosity", American Mathematical Society Translations, Series 2, 33(1963), pp. 285-290.

[22] F.M. ORR and C.M. JENSEN, Interpretation of pressure-composition diagrams for CO_2-crude oil systems, SPE 11125, presented at the 57th Annual Fall Technical Conference and Exhibition of the Society of Petroleum Engineers of AIME, New Orleans, September 26-29, 1982.

[23] D.-Y. PENG and D.B. ROBINSON, A new two-constant equation of state, Ind. Eng. Chem. Fundam. 15(1976), pp. 53-64.

[24] P.M. PRENTER, Splines and Variational Methods, John Wiley and Sons, New York, 1975.

[25] J.H. PREVOST, Mechanics of continuous porous media, Int. J. Engrg. Sci. 18(1980), pp. 787-800.

[26] R.F. SINCOVEC, Generalized collocation methods for time-dependent, nonlinear boundary-value problems, Soc. Pet. Eng. J.(1977), pp. 345-352.

[27] J.B. TEMPLE, Global existence of a class of 2×2 nonlinear conservation laws with arbitrary Cauchy data, preprint, Department of Mathematical Physics, Rockefeller University, New York, 1981.

[28] H.J. WELGE, A simplified method for computing oil recovery by gas or water drive, Trans. AIME 195(1952), pp. 91-98.

HOMOGENIZATION METHOD APPLIED TO BEHAVIOR SIMULATION OF NATURALLY FRACTURED RESERVOIRS

ALAIN BOURGEAT*

Abstract. The behavior of fluids in naturally fractured porous media can often be modeled by replacing the primitive flow equations with a set of "homogenized" flow equations formulated in terms of global pressures P and saturations S. This paper examines a medium interlaced by a network of uniformly spaced fractures and derives a system of coupled quasilinear partial differential equations for P and S. The homogenization consists of an application of integral averaging techniques involving two length scales. Rigorous arguments demonstrate convergence in the weak sense of the averaged one-dimensional equations to the homogenized equations as the fracture spacing vanishes. These arguments extend to flows in two or three space dimensions.

1. Introduction. In recent years the behavior of heterogeneous reservoirs has received considerable attention. A commonly encountered heterogeneous system is a naturally fractured reservoir in which two distinct types of porosities and permeabilities occur in the same formation. A region containing finer pores may have low permeability and porosity. This is called the matrix. The remaining region may have high permeability and porosity. This latter region is the set of interconnecting fractures and fissures of the rock. Hence, discontinuities in porosity and permeability exist throughout the reservoir.

During the seventies simplified models were used to study the single-phase pressure behavior of fractured reservoirs and the effects of fractures on pressure build-up curves [1-7]. These models are in essence similar to the radial model with double porosity introduced by Warren and Root [3]. Simulation of an entire reservoir system with multiple phases further complicates the problem. More recently several authors [8-12] have developed a modeling approach to multiphase flows in fractured reservoirs. All these works deal with numerical simulation by finite differences or finite element methods.

*Centre de Mathematiques, INSA, 20 Av. A. Einstein, 69621 Villeurbanne, France, and Department of Mathematics, University of Wyoming, Laramie, WY, 82071, U.S.A.

Our approach to this complex problem has been to model the flow behavior by a system of equations in which porosity Φ and permeability Ψ are considered to be functions depending on the space variables and then to derive "homogenized equations." The equations of two-phase flow in porous media that we are considering have been introduced by Chavent in [16-17]. This new formulation happens to be general and applies to both miscible or immiscible fluids. Thus they can model such various applications as oil and gas flow in reservoirs, water and oil flow in dams or soils, etc. These equations are formulated in terms of only two variables, the "reduced" water saturation S and the "global" pressure P. They make up a system of two coupled quasi-linear partial differential equations, an elliptic equation for P and a parabolic diffusion-transport equation for S. In the case of immiscible fluids the diffusion term in the saturation equation degenerates (it vanishes when $S=0$ or $S=1$).

If we consider a network of uniformly spaced fractures, then Φ and Ψ are rapidly oscillating periodic functions. For example, in an aquifer fissured reservoir, the Ψ variations may be from 3×10^{-4} up to 10 and the Φ variations from 2×10^{-2} up to 1 [6]. We assume that there is no well in the region we study and that fractures do not intersect boundaries or wells. The size of blocks is assumed to be small compared to the size of reservoirs. Call the ratio of the period scale to the reservoir scale ε. We see that the functions Ψ and Φ depend on ε and consequently the equations associated with this heterogeneous reservoir have solutions depending on ε.

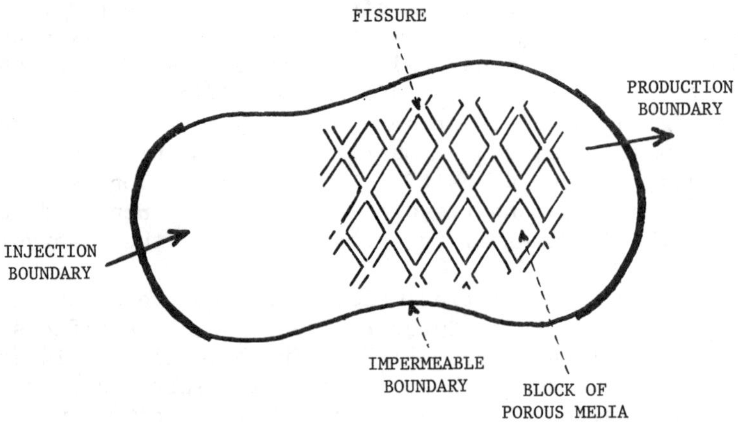

Figure 1. Schematic Fractured Reservoir

The "homogenized" equations are equations governing the asymptotic behavior of flows as ε tends to zero. They describe the flow in a "globally equivalent" homogeneous medium in which "effective" porosity and permeability depend no longer on ε or on the space variable. The procedure to derive "homogenized equations" is called "homogenization." In recent years mathematicians [13-15] have developed new methods for deriving them from the original complicated equations. Not very many mathematical works on homogenization have dealt with nonlinear partial differential systems and no general theory has been developed to homogenize nonlinear problems.

We shall proceed in two steps. The first step involves integral averaging over the period and two-length scale expansion. The "fast scale" accounts for the oscillations within one period and the "slow scale" accounts for variations within all the domain. The purpose of homogenization is to eliminate the variations on the short scale. This step will be used as a heuristic way of providing us with asymptotic equations for P and S (homogenized equations). The second step will prove in a formal and rigorous way the convergence (only in a weak sense) of complicated equations to homogenized equations as ε tends to zero. This step involves tools of nonlinear functional analysis in Sobolev spaces, G and S-convergence developed by E. DeGiorgi and S. Spangnolo [19], [20] and convergence by "compensated compactness" developed by F. Murat and L. Tartar [21-23]. The results we obtain are valid not only for fractured porous media but more generally for all porous media with a periodic inhomogeneous structure (stratified reservoirs for example). In these cases our results are consistent with the various averaging methods classically used in reservoir simulators. In some special cases (one-dimensional domain, stratified media) porosity and permeability are the standard arithmetic and harmonic means, widely used as approximations in numerical two-phase simulators.

For simplicity of explanation we will consider a one-dimensional domain. The two or three-dimensional cases can be handled in the same way, although the calculations are a little bit more cumbersome.

2. Behavior Equations. We are assuming that the two fluids and the rock are incompressible and that gravity effects are negligible. Two physical phases are present, a wetting phase (for example water) and a nonwetting phase (for example oil). We will deal with the mathematical model of two-phase flows introduced by G. Chavent. These equations are formulated in terms of only two dependent variables, a "global" pressure P which is intermediate between the water and oil pressures, and a "reduced" wetting phase saturation S. The gradient of this global pressure produces a mean oil-water filtration velocity vector q. Auxiliary functions a, b, d, α are computed from the usual relative permeability and capillary pressure curves. The relative permeability and capillary pressures are assumed independent of the position x when expressed in terms of the reduced saturation. Then functions a, b, d, α depend explicitly upon S only.

A principal feature of this model is its applicability to both miscible and immiscible fluids with only a change of auxiliary functions. Also the model admits a variational formulation which allows rigorous numerical and mathematical studies.

In the one-dimensional case the domain Ω is the open interval $(0,1)$, $\{0\}$ is the injection boundary, and $\{1\}$ the production boundary. The equations are:

Global Pressure Equation (Darcy's Law)

$$-\Psi^{\varepsilon}(x)\, d(S_{\varepsilon}) \frac{\partial P_{\varepsilon}(x;t)}{\partial x} = q(t) \qquad \text{in} \quad Q = (0,1) \times (0,T)$$

(1)

$$P_{\varepsilon}(1;t) = 0 \qquad t \in (0,T) \qquad \text{(pressure on the production boundary)}$$

Reduced Wetting Phase Saturation Equation (Mass Conservation Law)

$$\Phi^{\varepsilon}(x) \frac{\partial S_{\varepsilon}(x;t)}{\partial t} + \frac{\partial}{\partial x}\Big((q(t)b(S_{\varepsilon})\Big) - \frac{\partial}{\partial x}\left(\Psi^{\varepsilon}(x) \frac{\partial \alpha(S_{\varepsilon})}{\partial x}\right) = 0 \qquad \text{in} \quad Q$$

(2)

$$S_{\varepsilon}(1;t) = S_{\varepsilon}(0;t) = 0 \qquad \text{a.e. on} \quad (0,T)$$

$$S_{\varepsilon}(x;0) = S_{0}(x) \qquad \text{in} \quad \Omega$$

The boundary conditions in Equations (1) and (2) are not the more common conditions. They have been taken to allow further simplicity in mathematical discussions. Results will be the same with any boundary conditions of Dirichlet or Neumann or mixed type.

Subscripts ε are used to emphasize the dependence on ε of the terms in these equations. ε measures the small scale on which Ψ and Φ vary. As a consequence of the variation of Ψ and Φ, S and P also vary in this small scale. Our goal is to eliminate this small scale variation of S and P by considering the limits of S and P as ε tends to zero.

The mean flow q does not depend on the space variable x in the one-dimensional case and is equal to the mean flow given on the injection boundary $q(t) \equiv q(0;t)$. In multidimensional cases the mean flow is a vector and depends on both x and t. Equations (1) turn into $\mathrm{Div}(\vec{q}) = 0$.

$\alpha(S)$, the capillary diffusion, is a monotone increasing function, with $\alpha'(0) = \alpha'(1) = 0$ in immiscible cases. Then $a(S) = \int_0^S \alpha(s)\,ds$ vanishes for $S = 0$ and $S = 1$ in immiscible cases.

$b(S)$, the water transport caused by the mean filtration velocity q, is monotone increasing with $b(0) = 0$ and $b(1) = +1$.

$d(S)$, the "global" mobility, is strictly positive with $d(S) \geq d > 0$ in Ω.

All these functions are assumed to be continuous and bounded functions of $\mathbb{R}$ into $\mathbb{R}$.

Ψ^{ε} and Φ^{ε} are assumed to be bounded independently of ε and strictly positive in Ω.

The variational version of Equations (1) and (2) is:

$$\int_\Omega \Psi^\varepsilon(x)\, d(S_\varepsilon) \frac{\partial P_\varepsilon}{\partial x} \frac{dw}{dx}\, dx = [q(0;t)w(0)] \quad \text{a.e. on} \quad (0,T) \quad \text{for all} \quad w \in W$$

(3)

$$\Phi^\varepsilon(x) \left(\frac{\partial S_\varepsilon}{\partial t}, v\right) + \int_\Omega \Psi^\varepsilon(x) \frac{\partial \alpha(S_\varepsilon)}{\partial x} \frac{dv}{dx}\, dx - \int_\Omega b(S_\varepsilon) q(t) \frac{dv}{dx}\, dx = 0$$

$$\text{a.e. on} \quad (0,T) \quad \text{for all} \quad v \in V$$

W is the space of functions which belong to the Sobolev space $H^1(\Omega)$ and vanish on the production boundary. V is the space of functions which belong to the Sobolev space $H^1(\Omega)$ and vanish on both the production and injection boundary. V is the standard space denoted $H_0^1(\Omega)$. $\left(\, , \, \right)$ denotes the duality product between V and its dual space V'.

Equations (3) yield the following theorem of existence [24], [25]:

Theorem 1. There exist ("weak") solutions S_ε, P_ε of (3) such that

$$\text{(i)} \begin{cases} S_\varepsilon \in L^\infty(Q) & (S_\varepsilon \text{ is essentially bounded in } Q) \\ 0 \le S_\varepsilon \le 1 & \text{a.e. in } Q \end{cases}$$

$$\text{(ii)} \quad \beta(S_\varepsilon) \in L^2(0,T;V) \qquad (\beta(S_\varepsilon) \text{ is square integrable V valued})$$

(4) $$\text{(iii)} \quad P_\varepsilon \in L^2(0,T;W)$$

$$\text{(iv)} \quad \frac{\partial S_\varepsilon}{\partial t} \in L^2(0,T;V')$$

(v) in the miscible case only, moreover: $S_\varepsilon \in L^2(0,T;V)$

β^{-1}, the inverse of $\beta(S) = \int_0^S (a(s))^{1/2} ds$, is assumed to be a Hölder function of order $\theta \in (0,1)$; $s_0 \in L^2(\Omega)$ (square integrable on the domain Ω), $0 \le s_0(x) \le 1$; and $q(t) \in L^\infty(0,T)$ (essentially bounded on the interval $(0,T)$).

Remark 1. This theorem does not give uniqueness. The uniqueness can be got in special cases only (either one-dimensional domain or special values of auxiliary functions d and b) [24], [25].

3. Asymptotic Expansion. The essential difficulty in studying the asymptotic behavior of equations (1) and (2) is that a periodic L^P function converges toward its integral average only weakly and the limit of the product of two weakly convergent functions need not equal the product of the limits. Thereby we cannot guess "a priori" the limit of each of the terms in these equations.

By the change of variables $(x=x,\ y=x/\varepsilon)$ the functions $\Psi(y) := \Psi\left(\frac{x}{\varepsilon}\right) := \Psi^{\varepsilon}(x)$ and $\Phi(y) := \Phi\left(\frac{x}{\varepsilon}\right) := \Phi^{\varepsilon}(x)$ then are Y-periodic in the variable y. Ψ^{ε} and Φ^{ε} are εY-periodic in x. Moreover, the operator $\frac{\partial\cdot}{\partial x}$ becomes $\frac{\partial\cdot}{\partial x} + \frac{1}{\varepsilon}\frac{\partial\cdot}{\partial y}$.

y is measuring variations within one period cell (fast scale) and x variations within all the region (slow scale). Then we recall the following property:

> As $\varepsilon \to 0$, a Y-periodic function $F(y) := F\left(\frac{x}{\varepsilon}\right)$ belonging to $L^{\infty}(Y)$ converges to a constant $\tilde{F}$, called its average, $\tilde{F} = \frac{1}{\operatorname{meas}(Y)}\int_Y F(y)\,dy$, in the weak-star topology for $L^{\infty}(\Omega)$ (i.e. for all functions $\phi(x)$ integrable on Ω, $\int_{\Omega} F\left(\frac{x}{\varepsilon}\right)\phi(x)\,dx \to \tilde{F}\int_{\Omega}\phi(x)\,dx$).

The operator $\tilde{\cdot}$ will denote mean value in a period Y and will be called averaging.

This section is essentially heuristic. That is, we will not prove the convergence of expansions. To make complete rigorous sense of the procedure and to obtain a proof of convergence would require more smoothness of the functions involved than is physically reasonable. This section will only provide asymptotic equations; convergence will be proved rigorously without such strong smoothness in the next sections.

We are looking for an asymptotic expansion of P_{ε}, S_{ε} in the form:

$$\begin{aligned} P_{\varepsilon}(x;t) &= P_0(x;t) + \varepsilon P_1(x,y;t) + \varepsilon^2 P_2(x,y;t) + \cdots \\ S_{\varepsilon}(x;t) &= S_0(x;t) + \varepsilon S_1(x,y;t) + \varepsilon^2 S_2(x,y;t) + \cdots \end{aligned} \tag{5}$$

$$P_i,\ S_i \quad \text{Y-periodic in } y$$

Let us denote

$$
(6)\qquad
\begin{aligned}
Q^{\varepsilon}(x,y;t) &= -\Psi(y)d(S_{\varepsilon})\frac{\partial P_{\varepsilon}}{\partial x} = Q^{0}(x,y;t) + \varepsilon Q^{1}(x,y;t) + \cdots \\
R^{\varepsilon}(x,y;t) &= \Psi(y)\frac{d\alpha(S_{\varepsilon})}{dS}\frac{\partial S_{\varepsilon}}{\partial x} = R^{0}(x,y;t) + \varepsilon R^{1}(x,y;t) + \cdots
\end{aligned}
$$

Then Equations (1) and (2) become:

$$
(7)\qquad
\begin{aligned}
&\text{(i)}\quad \left(\frac{\partial}{\partial x}+\frac{1}{\varepsilon}\frac{\partial}{\partial y}\right)(Q^{\varepsilon}(x,y;t)) = 0 \\
&\text{(ii)}\quad \Phi(y)\frac{\partial}{\partial t}(S_0+\varepsilon S_1+\cdots) + q(t)\left[\frac{db(S_0)}{dS}+\varepsilon k\right]\left(\frac{\partial}{\partial x}+\frac{1}{\varepsilon}\frac{\partial}{\partial y}\right)(S_0+\varepsilon S_1+\cdots) \\
&\qquad - \left(\frac{\partial}{\partial x}+\frac{1}{\varepsilon}\frac{\partial}{\partial y}\right)(R^{0}+\varepsilon R^{1}+\cdots) = 0 \\
&\text{(iii)}\quad \text{Boundary Conditions and Initial Condition}
\end{aligned}
$$

Upon substituting (6) into (7) and equating like powers of ε, we obtain:

$$
(8)\qquad
\begin{aligned}
-\Psi(y)d(S_0)\left(\frac{\partial P_0}{\partial x}+\frac{\partial P_1}{\partial y}\right) &= Q^{0}(x,y;t) \\
\Psi(y)\frac{d\alpha(S_0)}{dS}\left(\frac{\partial S_0}{\partial x}+\frac{\partial S_1}{\partial y}\right) &= R^{0}(x,y;t)
\end{aligned}
$$

and

$$
(9)\qquad
\begin{aligned}
&\text{(i)}\quad \frac{\partial Q^{0}}{\partial y} = 0\ ,\quad \frac{\partial R^{0}}{\partial y} = 0 \quad (\text{"microscopic" equation}) \\
&\text{(ii)}\quad \frac{\partial}{\partial x}\left[\widetilde{Q^{0}}\right] = 0\ ,\quad [\tilde{\Phi}]\frac{\partial S_0}{\partial t} + q(t)\frac{db(S_0)}{dS}\frac{\partial S_0}{\partial x} - \frac{\partial}{\partial x}\left[\widetilde{R^{0}}\right] = 0 \quad (\text{"macroscopic" equations}) \\
&\qquad \text{B.C. and I.C.}
\end{aligned}
$$

The "microscopic" equations (9)-(i) are ordinary differential equations in the variable y and x is a parameter. P_1 and S_1 are unknown functions.

In the one-dimensional case we can solve these equations and after substituting in (9)-(ii) the "macroscopic" equations reduce to:

$$
(10)\qquad
\begin{aligned}
-\Psi^{\#}d(S_0)\frac{\partial P_0}{\partial x} &= q(t) \\
P_0(1;t) &= 0
\end{aligned}
$$

and

$$\Phi^{\#}\frac{\partial S_0}{\partial t}+\frac{\partial}{\partial x}[q(t)b(S_0)]-\Psi^{\#}\frac{\partial}{\partial x}\left[\frac{\partial\alpha(S_0)}{\partial x}\right]=0$$

(11) $$S_0(0;t)=S_0(1;t)=0$$

$$S_0(x;0)=\underset{\sim}{s}_0(x)$$

with $\Phi^{\#}=[\tilde{\Phi}(y)]$ and $\Psi^{\#}=\tilde{\tilde{\Psi}}:=1/\left[\frac{1}{\Psi(y)}\right]$ ($\tilde{\tilde{\Psi}}$ denotes the harmonic mean of $\Psi(y)$ on a period Y).

Thus (S_0,P_0) satisfy the same equations as do $(S_\varepsilon,P_\varepsilon)$ with $(\Phi^\varepsilon,\Psi^\varepsilon)$ replaced by the constants $(\Phi^{\#},\Psi^{\#})$. These equations (10), (11) are equations of two-phase flow in homogeneous media with porosity $\Phi^{\#}$ and permeability $\Psi^{\#}$. This justifies the names of "effective" porosity and permeability.

4. Convergence to Asymptotic Equations. We will not need more assumptions than we required to prove Theorem 1, except Φ^ε and Ψ^ε will be assumed bounded independently of ε.

Let us now consider $v\in L^2(0,T;V)$ and $w\in L^2(0,T;W)$ in the variational equations (3) and integrate over $(0,T)$. We obtain a priori estimates on $(P_\varepsilon,S_\varepsilon)$ and so we can extract a subsequence, still denoted $(S_\varepsilon,P_\varepsilon)$ which converges to (S,P) such that:

<u>Lemma 1</u>.

(i) $B_\varepsilon=\beta(S_\varepsilon)\rightharpoonup B$ in $L^2(0,T;V)$ weakly

(ii) $P_\varepsilon\rightharpoonup P$ in $L^2(0,T;W)$ weakly

(12)

(iii) $\Gamma_\varepsilon:=\Phi^\varepsilon\frac{\partial S_\varepsilon}{\partial t}$ and $\Delta_\varepsilon:=\frac{\partial}{\partial x}(b(S_\varepsilon)q(t))$ tend to Γ and Δ in $L^2(0,T;V')$ weakly

(iv) $\lambda_\varepsilon:=\frac{\partial\alpha(S_\varepsilon)}{\partial x}$, $\mu_\varepsilon:=b(S_\varepsilon)q(t)$ and $\nu_\varepsilon:=d(S_\varepsilon)\frac{\partial P_\varepsilon}{\partial x}$ tend to λ, μ, ν in $L^2(Q)$ weakly

(v) $S_\varepsilon\rightharpoonup S$ in $L^2(0,T;V)$ weakly (in the miscible case)

We now have to seek the values of Γ, Δ, λ, μ, ν because a priori we do not know these limits (see Remark 1). Using our hypothesis about β^{-1} (Hölder function of order θ, $\theta\in(0,1)$) and also using compactness of the injection of $W^{\theta s,2/\theta}(\Omega)$ in $L^{2/\theta}$ for all $s\in(0,1)$, we then have:

Lemma 2.

$$
(13) \qquad \begin{aligned} & S_\varepsilon \to S \quad \text{in} \quad L^2(\Omega) \\ & \Gamma = [\tilde{\Phi}]\frac{\partial S}{\partial t} \end{aligned}
$$

Furthermore the Lebesgue Theorem and properties (12), (13) yield:

$$
(14) \qquad \lambda = \frac{\partial \alpha(S)}{\partial x}, \quad \mu = b(S)q(t), \quad \nu = d(S)\frac{\partial P}{\partial x}, \quad \Delta = \frac{\partial}{\partial x}(b(S)q(t))
$$

Finally, after these two lemmas, we have to seek the limit of $\Theta_\varepsilon = \Psi^\varepsilon \nu_\varepsilon$ and $K_\varepsilon = \Psi_\varepsilon \lambda_\varepsilon$. To find the equations satisfied by these limits, we need to deal with limits in the $L^2(\Omega)$-valued distribution space $\mathcal{D}'(0,T;L^2(\Omega))$. Hence, we will use test functions ϕ that are scalar functions infinitely differentiable with support in $(0,T)$ $(C_0^\infty(0,T))$. Upon calling $\Theta_\varepsilon(\phi) := \int_0^T \Psi^\varepsilon(x)\nu_\varepsilon(x;t)\phi(t)dt$ and $\nu_\varepsilon(\phi) := \int_0^T \nu_\varepsilon(x;t)\phi(t)dt$, by using equation (1), we obtain:

$$
(15) \qquad \begin{aligned} & \text{(i)} \quad \frac{\partial \Theta_\varepsilon(\phi)}{\partial x} = 0 \quad \text{and hence} \quad \Theta_\varepsilon(\phi) \to \Theta(\phi) \quad \text{in} \quad L^2(Q) \\ & \text{(ii)} \quad \nu_\varepsilon(\phi) \rightharpoonup \nu(\phi) = \widetilde{\left[\frac{1}{\Psi}\right]}\Theta(\phi) \end{aligned}
$$

In view of (15)(ii), which happens for all $\phi \in C_0^\infty(0,T)$, we get:

Theorem 2. P is a solution of

$$
(16) \qquad \begin{aligned} & -\approx{\Psi} d(S)\frac{\partial P}{\partial x} = q(t) \\ & P(1;t) = 0 \end{aligned}
$$

In the same way as for P, calling $\alpha_\varepsilon(\phi) = \int_0^T \alpha(S_\varepsilon)\phi(t)dt$ and $\mu_\varepsilon(\phi) = \int_0^T \mu_\varepsilon(x;t)\phi(t)dt$, and invoking equation (2) gives:

$$
(17) \qquad \frac{\partial}{\partial x}\left(\Psi^\varepsilon \frac{\partial \alpha_\varepsilon(\phi)}{\partial x}\right) = \frac{d\mu_\varepsilon(\phi)}{dx} - \int_0^T \Phi^\varepsilon S_\varepsilon \frac{d\phi}{dt}\, dt
$$

Equation (17) is a standard linear elliptic equation [17] with a right hand side which converges in V' strongly to $\frac{d\mu(\phi)}{dx} + [\tilde{\Phi}] \int_0^T S \frac{d\phi}{dt}\, dt$ from (12)(iv), (13) and (14). So we know the left hand side tends to $\approx{\Psi} \frac{\partial}{\partial x}\left(\frac{\partial \alpha(\phi)}{\partial x}\right)$ for all test functions ϕ. The above proves:

Theorem 3. S is a solution of

$$(18)\qquad \begin{aligned} &[\tilde{\Phi}]\frac{\partial S}{\partial t} + \frac{\partial}{\partial x}[q(t)b(S)] - \tilde{\tilde{\Psi}}\,\frac{\partial}{\partial x}\left[\frac{\partial \alpha(S)}{\partial x}\right] = 0 \\ &S(0;t) = S(1;t) = 0 \\ &S(x;0) = s_0(t) \end{aligned}$$

Remark 2. The theorems (3) and (4) give the same results we obtain by asymptotic expansion in (10) and (11).

5. Conclusion. There is no difficulty in extending the above results to a 2- or 3-dimensional model because we used only properties valid for multidimensional cases. We did not need or use specific one-dimensional regularities or properties. The only difference is that we obtain partial differential equations instead of ordinary differential equations (9)(i) for finding the homogenized coefficients. Consequently the proof of convergence is a little bit more cumbersome.

In the multidimensional case the homogenized tensor $\Psi^{\#}_{i\ell} = \left[\Psi_{i\ell} + \Psi_{ij}\frac{\partial w^{\ell}}{\partial y_j}\right]$ where $w^{\ell}(y)$ is the Y-periodic solution of

$$-\frac{\partial}{\partial y_i}\left[\Psi_{ij}\frac{\partial w^{\ell}}{\partial y_j}\right] = \left[\frac{\partial \Psi_{i\ell}}{\partial y_i}\right]$$

The homogenized equations are then:

$$-\frac{\partial}{\partial x_i}\left[\Psi^{\#}_{i\ell}\,d(S)\frac{\partial P}{\partial x_\ell}\right] = 0$$

$$\tilde{\Phi}\,\frac{\partial S}{\partial t} + \frac{\partial}{\partial x_i}\left[b(S)\Psi^{\#}_{i\ell}\,d(S)\frac{\partial P}{\partial x_\ell}\right] - \frac{\partial}{\partial x_i}\left[\Psi^{\#}_{i\ell}\,\frac{\partial \alpha(S)}{\partial x_\ell}\right] = 0$$

After homogenization isotropy is no longer generally valid. That is, the diagonal terms of the matrix $\Psi^{\#}$ are not necessarily equal. For example, in case of stratified media with strata orthogonal to the direction x_3, we obtain [22] the homogenized permeability tensor

$$\Psi^{\#} = \begin{bmatrix} \tilde{\Psi}_{11} & 0 & 0 \\ 0 & \tilde{\Psi}_{22} & 0 \\ 0 & 0 & \tilde{\tilde{\Psi}}_{33} \end{bmatrix}$$

More generally, in a more complex medium, the homogenized matrix $\Psi^{\#}$ can have nonzero off-diagonal terms.

Likewise, there is no major difficulty in extending the results to cases where permeability and porosity are dependent not only on x/ε but depend on space variables too. This is a problem of nonuniform homogenization. Φ and Ψ are functions of both x and x/ε. The results are the same but the proof of convergence is more sophisticated because we must discretize the operators with respect to x and prove the convergence for these discretized operators.

All the results obtained above are evidently valid more generally for all types of reservoirs with periodic inhomogeneous media.

Acknowledgement. This research has been partially supported by the Special Energy Year at the University of Wyoming. I thank K. Gross for the invitation to take part in the Special Energy Year and R. Ewing for useful discussions.

REFERENCES

[1] J.E. WARREN and H.S. PRICE, Flow in heterogeneous porous media, Trans. Soc. Pet. Eng. of AIME 222(1961), pp. 153-167.

[2] A.S. ODEH, Unsteady state behavior of naturally fractured reservoirs, Soc. Pet. Eng. J. (March 1965), pp. 60-66.

[3] J.E. WARREN and P.J. ROOT, The behavior of naturally fractured reservoirs, Soc. Pet. Eng. J. (September 1963), pp. 245-255.

[4] H. KAZEMI, Pressure transient analysis of naturally fractured reservoirs with uniform fracture distribution, Soc. Pet. Eng. J. (December 1969), pp. 451-462.

[5] H. KAZEMI, M.S. SETH, and G.W. THOMAS, The interpretation of interference tests in naturally fractured reservoirs with uniform fracture distribution, Soc. Pet. Eng. J. (December 1969), pp. 463-470.

[6] P.J. CLOSMANN, An aquifer model for fissured reservoirs, Soc. Pet. Eng. J. 15(5) (1975), pp. 385-392.

[7] G.D. CRAWFORD, A.R. HAGEDORN, and A.E. PIERCE, Analysis of pressure buildup tests in a naturally fractured reservoir, J. Pet. Tech. (November 1976), pp. 1295-1300.

[8] A.B. GUREGHIAN, A study by the finite element method of the influence of fractures in confined aquifers, Soc. Pet. Eng. J. 15(2) (1975), pp. 181-191.

[9] R.H. ROSSEN, Simulation of naturally fractured reservoirs with semi-implicit source terms, Soc. Pet. Eng. J. (June 1977), pp. 201-210.

[10] A. ASFARI and P.A. WITHERSPOON, Numerical simulation of naturally fractured reservoirs, Paper SPE 4290 presented at the SPE-AIME Third Symposium on Numerical Simulation of Reservoir Performance, Houston, January 11-12, 1973.

[11] J.F. ABEL and J.O. DUGUID, Finite element Galerkin method for analysis of flow in fractured porous media in Finite Element Methods, U.A.H. Press, 1974, pp. 599-615.

[12] J. LEFEBRE DUPREY and L.M. WEILL, Frontal displacement in a fractured reservoir in Finite Element Methods, U.A.H. Press, 1974, pp. 653-671.

[13] J.B. KELLER, Darcy law for flow in porous media and the two-space method in Nonlinear Partial Differential Equations in Engineering and Applied Science, R.L. Sternberg, A.J. Kalinowski, and J.S. Papadakis, eds., Marcel Dekker, New York, 1980, pp. 429-443.

[14] A. BENSOUSSAN, J.L. LIONS, and G. PAPANICOLAOU, Asymptotic Analysis for Periodic Structures, North Holland, Amsterdam, 1978.

[15] E. SANCHEZ PALENCIA, Non-homogeneous media and vibration theory, Lecture Notes in Physics, 127, Springer-Verlag, Berlin, 1980.

[16] G. CHAVENT, A new formulation of diphasic incompressible flows in porous media, Lecture Notes in Mathematics, 503, Springer Verlag, Berlin, 1976, pp. 258-270.

[17] G. CHAVENT, The global pressure: a new concept for the modelization of compressible two-phase flows in porous media, Proceedings of Euromech, 143, Delft, September 2-4, 1981.

[18] G. CHAVENT, Mathematical modeling of mono and diphasic oil or gas reservoirs, La Recuperation du Petrole: Modeles Mathematiques et Numeriques, INRIA, B.P. 105, F. 78150, Le Chesnay, October 25-29, 1982, pp. 45-118.

[19] E. DE GIORGI, Convergence problems for functionals and operators, Proceedings of the International Meeting on Recent Methods in Nonlinear Analysis, E. De Giorgi, E. Magenes, and V. Mosco, eds., Pittagora, Bologna, 1978, pp. 131-188.

[20] S. SPAGNOLO, Some convergence problems, Conv. Ist. Sup. Alta Mat. Roma March 1974, 18(1976), pp. 391-398.

[21] F. MURAT, Compacite par compensation II, Proceedings of the International Meeting on Recent Methods in Nonlinear Analysis, E. De Giorgi, E. Magenes, and V. Mosco, eds., Pittagora, Bologna, 1978, pp. 245-255.

[22] L. TARTAR, Cours Peccot, College de France, Paris, 1977.

[23] B. DACOROGNA, Weak continuity and weak lower semicontinuity of nonlinear functionals, Lecture Notes in Mathematics, 922, Springer Verlag, Berlin, 1982.

[24] G. CHAVENT, About the identification and modeling of miscible or immiscible displacement in porous media, IFIP Working Conference on Distributed Parameter Systems Modeling and Identification, Lectures Notes in Control and Information Science, 1, Springer Verlag, Berlin, 1978, pp. 196-220.

[25] G. GAGNEUX, Deplacements de fluides non-miscibles incompressibles dans un cylindre poreux, Journal de Mecanique 19(2) (1980), pp. 295-325.

VISCOUS FINGERING IN HYDROCARBON RECOVERY PROCESSES***

RICHARD E. EWING* AND JOHN H. GEORGE**

Abstract. The understanding of the viscous fingering phenomenon in porous media flow and the incorporation of the effects of fingering in the numerical simulation of enhanced oil recovery processes is essential for optimization of hydrocarbon recovery. In this paper, the literature on viscous fingering instabilities is surveyed and an attempt is made to describe the current understanding of the related causes and effects. Both miscible and immiscible flow regimes are discussed and compared. Simplified models of the process, such as the Hele-Shaw cell, are considered and the results are interpreted via linear and weakly nonlinear perturbation methods and dimensional analysis. Avenues for incorporating more nonlinear effects to describe the finger growth are presented. Directions for further research are suggested.

1. Introduction. Primary recovery techniques, where the resident formation pressures provide the primary energy to extract the hydrocarbons, often leave 70-90% or more of the hydrocarbons in the reservoir [5,6,28,48,54,55,62,97]. Subsequently, water or other fluids may be injected into some wells in the reservoir while petroleum is produced through others. This serves the dual purpose of maintaining high formation pressures and flow rates and also of flooding the porous medium to physically displace some of the oil and push it toward the production wells. This type of pressure maintenance and water-flooding is usually termed secondary recovery. In this process the water does not mix with the oil due to surface tension effects and the process is termed immiscible displacement. Often, water-flooding is not sufficiently effective and gases like CO_2 are injected to interact with the petroleum to try to form a single, flowing phase, and reduce the surface tension effects which trap the residual petroleum in the small pores in the rock and allow greater sweep efficiency. A process where significant mixing of this type between the fluids occurs is termed miscible displacement. Miscible displacement and other tertiary processes, such as steam-flooding, in-situ combustion, and use of polymers or surfactants are described in somewhat more detail in [5,8,16,23,26,32,35,36,39,51,93,97,105].

In both miscible and immiscible displacements, the process of pushing a heavy, viscous oil through a porous medium with a lighter, less viscous fluid can be a very unstable one. If the flow rate is sufficiently high, the interface between the resident

*Departments of Mathematics and Petroleum Engineering and the Center for Enhanced Oil Recovery, University of Wyoming, Laramie, WY 82071

**Department of Mathematics, University of Wyoming, Laramie, WY 82071

***This research was supported in part by a Contract No. DAAC29-84-K-0002 from the U.S. Army Research Office.

petroleum and the invading fluid becomes unstable and tends to form long fingers which grow in length toward the production wells, bypassing much of the hydrocarbons. Once a path consisting of the injected fluid has extended from an injection well to a production well, that production well will henceforth produce primarily the injection fluid which flows more easily due to its lower viscosity and higher mobility. The production of petroleum from that well is then greatly reduced if not essentially stopped. This phenomenon, termed viscous fingering, is well known [3,4,7,23,28,33,38,43,47,48,52,57, 59,60,62,63,64,66,71,72,74,76,77,80,84,87,88,90,91,104,110,112] and is a serious problem in hydrocarbon recovery. This problem and different techniques for understanding and modeling it are the subjects of this paper.

Analysis must be applied to the equations which describe the microscopic physics of multiphase flow to understand the conditions under which small flow perturbations caused by tortuous flow in the porous media will become unstable and will grow into large fingers which affect the flow on a macroscopic level. The understanding of this instability is crucial, both in attempts to stabilize the flow via polymers, etc., and to model the fingering phenomenon when it cannot be controlled. There are three distinct problems associated with the viscous fingering phenomenon. Knowledge of the condition causing the onset of the instability is one major goal. Then the understanding of the nonlinear effects which cause certain fingers to grow preferentially and coalesce to form larger fingers is essential. The rate of this finger growth must also be determined. Once the growth of the fingers is understood, from a microscopic to a macroscopic level, statistical methods should be used to incorporate the large-scale effects of fingering into the mathematical models used to simulate the field scale process without trying to treat the individual fingers.

The objectives of this paper are to present an overview of the problem of viscous fingering pointing out key ideas and outstanding problems. A survey of the extensive literature on various aspects of the problem will be given emphasizing the major contributions to our present understanding. Then we shall discuss directions for future research and promising approaches aimed at understanding the phenomena sufficiently well to accurately predict the oil yield under various production strategies and ultimately to maximize the hydrocarbon extraction.

2. Specific Assumptions and a Model Problem. There are many different types of fingering phenomena which arise in the various enhanced recovery processes. Although there are important differences in the stability analyses of these phenomena, for expositional purposes we shall make certain simplifying assumptions and focus on a specific model problem which is sufficiently general to illustrate the basic problems involved in the stability analyses. We shall first neglect any chemical or thermal effects and shall initially assume that we have an immiscible displacement process. Extensions to miscible displacement are discussed briefly in Section 3 and in [25,26,38,44,71,72,73,74]. We shall only consider linear or weakly nonlinear variations from a "flat" water-oil interface [4,7,19,26,27,28,33,38,42,48,49,51,54,55,59,63,64,66,74,75,76,85,92,93,104,108, 110,111]. We also treat only linear end-to-end drives in a homogeneous, two-dimensional reservoir in this discussion. Discussion of radial flow effects can be found in [1,23,24,48,101].

If we assume an initial, linear interface between the two fluids, the tortuous flow through the porous media at the microscopic level will introduce small "bumps" or irregularities in the interface. The major questions to be addressed are: (1) under what conditions will the perturbations grow into fingers; and (2) how will the fingers grow. We shall first consider the conditions for onset of instability.

The assumption of immiscibility implies that a sharp interface is maintained between the two fluids. Technically this never happens, since the interface is at least several molecular diameters thick, depending on the degree of mixing [48,59,81,82,99] and possible capillary pressure effects. We shall introduce the fingering concept with an extremely simple model* that includes many of the fundamental principles. Assume:

A1. A sharp interface exists between the water and oil.
A2. The pressure gradient is continuous across the interface.
A3. The flow is in the x-direction with velocity U.
A4. The finger is centered in the flow channel and is symmetric about the x-axis.

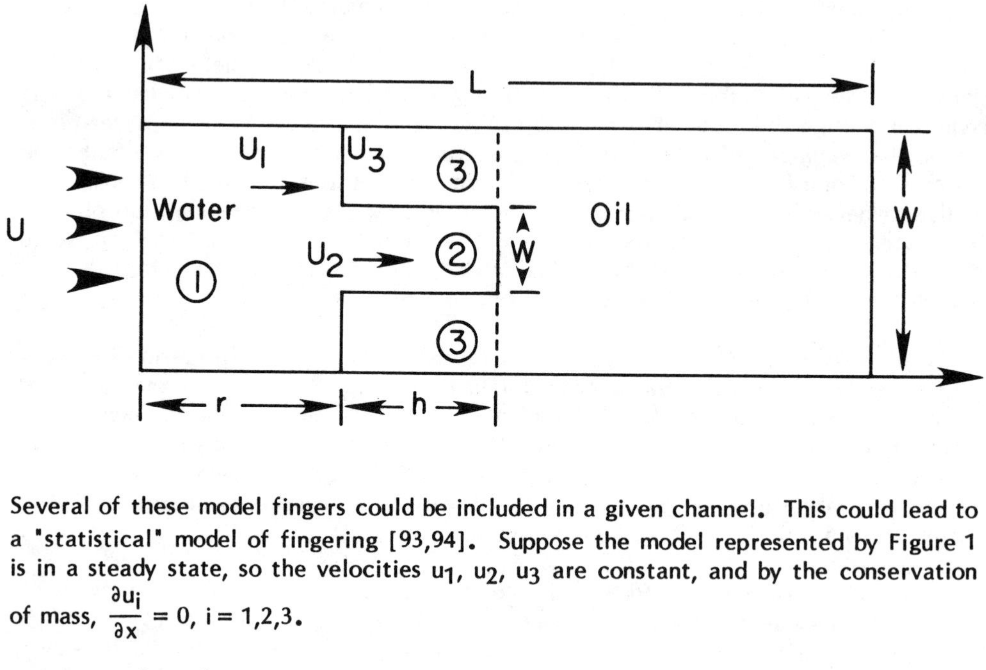

Several of these model fingers could be included in a given channel. This could lead to a "statistical" model of fingering [93,94]. Suppose the model represented by Figure 1 is in a steady state, so the velocities u_1, u_2, u_3 are constant, and by the conservation of mass, $\frac{\partial u_i}{\partial x} = 0$, $i = 1,2,3$.

Assume that the relationship between velocity and pressure gradient given by Darcy's Law is valid:

$$u_1 = -m_w \frac{\partial p_1}{\partial x}$$

(1)
$$u_2 = -m_w \frac{\partial p_2}{\partial x}$$

$$u_3 = -m_o \frac{\partial p_3}{\partial x}$$

*These ideas follow a communication with Peter Purcell.

where the respective mobilities are

$$m_w = \frac{k_w}{\mu_w}, \; m_o = \frac{k_o}{\mu_o} ,$$

with

$$k_i = \text{relative permeability, } i = o,w$$

$$\mu_i = \text{viscosity, } i = o,w$$

and the <u>mobility</u> <u>ratio</u> is given by

$$m = \frac{m_w}{m_o} .$$

By assumption A2, at the finger,

(2) $$\frac{\partial p_2}{\partial x} = \frac{\partial p_3}{\partial x} ,$$

which by Darcy's Law implies

(3) $$u_2 = mu_3 .$$

We assume the finger is growing, or that $u_2 > u_3$ which, from (3), yields

(4) $$m > 1 .$$

This condition on the mobility ratio is clearly necessary for unstable flow. In this extremely simplified model this condition determines the onset and growth of the instabilities. We will present a more complete description of the stability problem later for different flow problems when other more complex conditions determine the onset and growth properties.

Let

(5) $$\omega = \frac{w}{W}$$

be the dimensionless finger width. By appealing to continuity of the velocities,

(6) $$U = u_1 = \omega u_2 + (1 - \omega)u_3 .$$

Then, combining (3) and (6), we see that

(7) $$u_3 = \frac{U}{1 + \omega(m - 1)} = \frac{u_2}{m} .$$

Assume the finger growth begins at some time $t = t_0 \geq 0$. Then

$$(8)\qquad \frac{dh}{dt} = \begin{cases} 0, & 0 \le t < t_0 \\ u_2 - u_3 = \dfrac{(m-1)U}{1+(m-1)\omega}, & t \ge t_0 \end{cases}$$

and, $h(t_0) = 0$. Solving (8) yields

$$(9)\qquad h = \begin{cases} 0, & 0 \le t < t_0 \\ \dfrac{(m-1)U(t-t_0)}{1+(m-1)\omega}, & t \ge t_0 . \end{cases}$$

Also,

$$(10)\qquad \frac{dr}{dt} = \begin{cases} U, & 0 \le t < t_0 \\ u_3 = U - \dfrac{\omega(m-1)U}{1+(m-1)\omega} = \dfrac{U}{1+(m-1)\omega}, & t \ge t_0 \end{cases}$$

with

$$r(t_0) = Ut_0 .$$

Integrating (10), we see that

$$(11)\qquad r = \begin{cases} Ut, & 0 \le t < t_0 \\ Ut - \omega h = U\left[t - \dfrac{\omega(m-1)(t-t_0)}{1+(m-1)\omega}\right], & t \ge t_0 . \end{cases}$$

The viscous energy in a uniformly saturated porous medium of volume V is given by [33,93]

$$E_{vis} = \frac{1}{m}\int_V |u|^2\, dV$$

where u is the velocity and m is the mobility ratio. By letting A_w be the area containing water and A_o the area containing oil, the viscous energy can be written for our problem as

$$(12)\qquad E_{vis} = \frac{1}{m_w}\int_{A_w} |u_w|^2\, dA_w + \frac{1}{m_o}\int_{A_o} |u_o|^2\, dA_o$$

$$(13)\qquad \simeq \frac{U^2 rW}{m_w} + \frac{hw}{m_w}\left[\frac{mU}{1+(m-1)\omega}\right]^2 + \frac{h(W-w)}{m_o}\left[\frac{U}{1+(m-1)\omega}\right]^2 + \frac{(L-r-h)W}{m_o}U^2 .$$

Since $m_w = m_o m$,

$$(14)\qquad E_{vis} = \frac{WU^2}{m_o}\left[\frac{r}{m} + \frac{h}{[1+(m-1)\omega]^2} + L - r - h\right] .$$

By definition of r and h in (9) and (11), we see that

$$
(15) \qquad E_{vis} = \begin{cases} \dfrac{WU^2}{m_o}\left[L - \dfrac{(m-1)Ut}{m}\right], & 0 < t < t_0 \\[2ex] \dfrac{WU^2}{m_o}\left[L - \dfrac{(m-1)Ut}{m} - \dfrac{(m-1)^3\omega(1-\omega)U(t-t_0)}{m[1+(m-1)\omega]}\right], & t \geq t_0 . \end{cases}
$$

When $m > 1$, E_{vis} decreases, i.e., $\dot{E}_{vis} < 0$, and when $t \geq t_0$, E_{vis} has a minimum $\left[\dfrac{\partial E_{vis}}{\partial \omega} = 0,\ \dfrac{\partial^2 E_{vis}}{\partial \omega^2} > 0\right]$ at

$$
(16) \qquad \omega = \frac{1}{m+1} .
$$

Intuitively, we expect the finger width which minimizes the dissipation of energy to be the preferred one. In a heuristic way, this example illustrates how the energy method determines stability [12,19,20,21,22].

The basic concept in using energy methods is analogous to the ideas of Liapunov stability theory. A positive energy-like function, such as the Helmholtz free energy, is used to compute the time derivative of the function along a solution. The derivative is bounded in such a way that stability is ensured in a neighborhood of the solution [11,12,19,20,21,22,55]. Dyson [22] has re-fined the energy method to include ideas from directed graphs and circuit analysis (i.e., stability on a finite set of possible equilibria). In order to extend these ideas in one of many possible ways [19,20,21,22,39,40], we need to introduce some basic concepts.

A contact line is a geometric curve formed when an interface between two immiscible fluids intersects a solid in three space dimensions. If V_i is the volume occupied by liquid i, $i = 1,2$, and S is the interface, $V_1 \cap V_2$, then the contact line (or curve) is $S \cap \partial(V_1 \cup V_2)$.

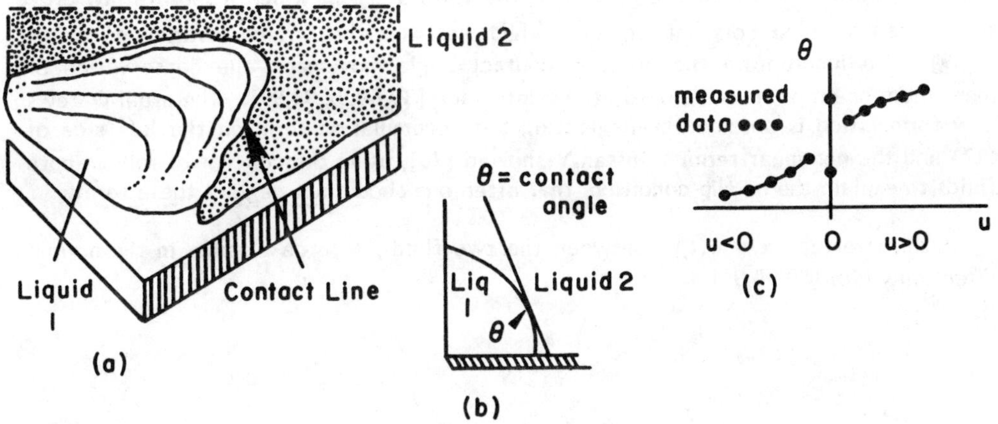

The contact angle is usually defined assuming the liquids are static [19], and is the angle made by the tangent to Liquid 1 at the contact line (see Figure 2b). The contact angle is obviously not constant, and can change as a function of velocity (hysteresis effects) (see Figure 2c), temperature (Marangoni effects), position, and other factors. The concept of contact angle is closely related to the idea of wettability used in the engineering literature [1,15,39,40,47,48,50,51,55]. Wettability is usually a function of the static contact angle. One major problem in the stability analysis of variable contact angle is how to interpret the "no slip" boundary condition at the solid surface. The best interpretation seems to be that of allowing convection patterns to occur in the two fluids so that the nonslip condition is still satisfied [20]. The energy method has been applied to several problems which allow movement of the contact angle, and has the promise of solving a wide class of dynamic contact angle problems [20].

3. Stability Theory. The equations used to describe fluid flow in porous media are obtained by applying various averaging techniques to the Navier-Stokes equations which describe the microscopic flow properties. The averaging is done over tubes of varying geometry and tortuosity [1,17,41,48,60,92,97,98]. Clearly, the Navier-Stokes equations do not describe the flow on a macroscopic level. In the averaging process, various important assumptions must be made. These assumptions greatly effect the resulting equations used to model the macroscopic flow.

A generalized Darcy's Law, obtained via this averaging process, can be written as [41,111]

$$\frac{\rho_i}{\varepsilon}\frac{\partial \underline{u_i}}{\partial t} = -\nabla(p_i + \rho_i q \underline{x}) - \frac{1}{m_i}(\underline{u_i} + \underline{U}) + b|\underline{u_i}|\underline{u_i} \tag{17}$$

where

$$\underline{u_i} = \begin{pmatrix} u_i \\ v_i \end{pmatrix}, \quad i = o,w ,$$

is the velocity of the $\underline{i^{th}}$ fluid, m_i and $\underline{U}$ are as before, ρ_i are the densities, p_i are the pressures, q is the acceleration due to gravity and ε and b are flow parameters. The nonlinear term is an average of terms of the form $u\frac{\partial u}{\partial x}$ in which a closure law produces the so-called Forscheimer term which satisfies the "frame invariance" axiom [1,39]. To incorporate the dynamic contact angle movement, the conservation of momentum equation must be used at the interface [39,60,97,98,99]. The usual Darcy's Law formulation is obtained by neglecting the accumulation term on the left side of (17) and the nonlinear term. Dussan V showed [20], using the concept of rolls in both fluids to explain the no slip condition, that often one should not neglect these terms.

At the interface $x = \eta(t,y)$ between the two fluids, a force balance in the normal direction yields [59,98,99]

(18)
$$p_w - p_o = T_e \eta_{yy}(1+\eta_y^2)^{-3/2} - 2\mu_o\left(\frac{\partial v_o}{\partial y} + 2\mu_w\frac{\partial v_w}{\partial y}\right)\left(\frac{\eta_y^2}{1+\eta_y^2}\right)$$
$$-2\left[\mu_o\left(\frac{\partial u_o}{\partial y}+\frac{\partial v_o}{\partial y}\right) - \mu_w\left(\frac{\partial u_w}{\partial y}+\frac{\partial v_w}{\partial y}\right)\right] - 2\left[\mu_o\frac{\partial v_o}{\partial y} - \mu_w\frac{\partial v_w}{\partial y}\right]\frac{1}{1+\eta_y^2} + P_c(t)$$

where μ_o, μ_w are the dynamic viscosities of oil and water respectively. These terms are usually neglected, along with the fact that T_e is not constant. If mass transfer is permitted, then other terms must be added [98,99] in (18).

By assuming both fluids incompressible the conservation of mass equation is

(19)
$$u_{i,x} + v_{i,y} = 0 \qquad\qquad i = o,w$$
$$u_i(t,\pm\infty,y) = 0$$

By letting Φ_i be a potential function and

(20)
$$\underline{u_i} = -\nabla\Phi_i ,$$

the conservation of mass becomes

(21)
$$\nabla^2\Phi_i = 0$$
$$\eta < x < \infty, \qquad\qquad i = o$$
$$-\infty < x < \eta, \qquad\qquad i = w$$
$$\Phi_o(\infty,y) = 0,\ \Phi_w(-\infty,y) = 0 .$$

The free surface condition is [98,99]

$$\frac{D}{Dt}(x - \eta(t,y)) = 0 ,$$

or

(22)
$$\frac{\partial\eta}{\partial t} - u_i + v_i\frac{\partial\eta}{\partial y} = \frac{\partial\eta}{\partial t} + \frac{\partial\Phi_i}{\partial x} - \frac{\partial\Phi_i}{\partial y}\frac{\partial\eta}{\partial x} = 0 .$$

To calculate the stability, first linearize all of the equations; second, from (20), replace $\underline{u_i}$ by $-\nabla\Phi_i$ in (17) and integrate to solve for p_i. Then substitute p_i into (18) and solve the result simultaneously with (21) and (22). Letting $\mu_o = \mu_w = 0$, the new equation (18) is

(23)
$$\frac{\rho_w}{\varepsilon}\frac{\partial\Phi_w}{\partial t} - \frac{\rho_o}{\varepsilon}\frac{\partial\Phi_o}{\partial t} - \eta\left[\rho_w - \rho_o - U\left(\frac{1}{m_w} - \frac{1}{m_o}\right)\right] - \frac{\Phi_w}{m_w} - \frac{\Phi_o}{m_o} = T_e\frac{\partial^2\eta}{\partial\eta^2} ,$$

and the boundary condition (22) becomes

(24) $$\frac{\partial \eta}{\partial t} = -\frac{\partial \Phi_o}{\partial x} = -\frac{\partial \Phi_w}{\partial x} .$$

A solution of (21), (23), and (24) is

(25) $$\eta = 0,\ \Phi_i = 0,\ i = o,w .$$

Perturbing from this steady state solution (25) by $\delta\eta$, $\delta\Phi_i$, i = o,w where

$$\delta\eta = \sum_{n=1}^{\infty} a_n S_n(y)$$

$$\delta\Phi_o = -\sum_{n=1}^{\infty} \frac{da_n(t)}{dt} \frac{e^{\alpha_n x}}{\alpha_n} S_n(y)$$

$$\delta\Phi_w = \sum_{n=1}^{\infty} \frac{da_n(t)}{dt} \frac{e^{-\alpha_n x}}{\alpha_n} S_n(y) .$$

Here the $S_n(y)$, n = 1,2,... satisfy the boundary conditions in (21). Let

$$a = \frac{\dfrac{\rho_o}{\varepsilon_o} + \dfrac{\rho_w}{\varepsilon_w}}{\dfrac{\mu_o}{k_o} + \dfrac{\mu_w}{k_w}}$$

and

$$b = \frac{\left(\dfrac{\mu_o}{k_o} - \dfrac{\mu_w}{k_w}\right) U + (\rho_o - \rho_w) g}{\dfrac{\mu_o}{k_o} + \dfrac{\mu_w}{k_w}} .$$

Then the stability parameter is given by [112]

(26) $$ar^2 + r - b = 0 .$$

If a is small,

$$r_1 = b - ab^2$$

$$r_2 = -\frac{1}{a} - b .$$

r = b is the Chuoke condition [7]. If b > 0, $a > \frac{1}{b}$, both roots are negative. This linear stability condition was obtained by Chuoke [7] and Saffman and Taylor [88] for the Hele-Shaw problem. The condition predicts only the onset of instability and cannot give information about the finger growth. Only by adding the nonlinear terms to the stability analysis can the character of the fingers be understood [64,67,87].

If mixing is allowed, as in miscible displacement, only the water equation is needed in (17) and the mobility ratio is a function of the water concentration C. The convection dispersion relation [23,70] is

(27)
$$\phi \frac{\partial C}{\partial t} + u_w \frac{\partial C}{\partial x} + v_w \frac{\partial C}{\partial y} = \nabla\cdot(K\nabla C) .$$

If a portion of the flow domain were to contain a mass of solute called a tracer, as the flow continues the tracer would gradually spread and occupy an increasing portion of the flow domain. The phenomenon is called hydrodynamic dispersion [70]. The dispersion is represented in equation (27) by a matrix [23,39,70] $K = (k_{ij})$ where

(28)
$$k_{ij} = I\, d_m \phi(\underline{x}) + \frac{d_l}{|\underline{u}_w|}\begin{pmatrix} u_w^2 & u_w v_w \\ u_w v_w & v_w^2 \end{pmatrix} + \frac{d_t}{|\underline{u}_w|}\begin{pmatrix} u_w^2 & -u_w v_w \\ -u_w v_w & v_w^2 \end{pmatrix}$$

and

$$d_m = \text{molecular diffusion coefficient}$$

$$d_l = \text{longitudinal dispersivity}$$

$$d_t = \text{transversal dispersivity}$$

in an isotropic medium. If bedding planes and other structural anomalies are present, the statistical averaging process used to obtain (28) would change the coefficients.

The boundary conditions for (27) are

$$C(t,x_1,y) = 1$$

$$C(t,x_2,y) = 0$$

$$C(0,x,y) = \begin{cases} 1, & x < 0 \\ 0, & x \geq 0 \end{cases}$$

where x_1 is the point where C is all water and $x_2 > x_1$ is the point where there is no water. Usually it is assumed that $x_1 = -\infty$, $x_2 = +\infty$ for ease of solution. A stability analysis for this special case has been developed [25,26,38,44,71,72,73,74], and develops in much the same way as the preceding stability argument. Simplified versions have been proposed and seem quite successful [10,26,33].

A complete understanding of the fingering phenomenon must include the nonlinear terms that "shape" the fingers. To include nonlinear effects, Outmans [63] has used asymptotic methods on the nonlinear equations. His calculations indicate that the nonlinear terms play a significant role in the behavior and growth of fingers. Nayfeh [59] has developed a "weakly nonlinear" theory for fingering in porous media flow. His theory assumes that the amplitude function $a_n(t)$, with n corresponding to b obtained from the Chuoke condition, is expanded in the form

(29)
$$\frac{da_n(t)}{dt} = \frac{dA}{dt} = bA + \varepsilon^2\sigma_2 A^3 + \cdots .$$

By using a Galerkin-type projection, Nayfeh obtained

(30)
$$\frac{dA}{dt} = \alpha A + F_{mn}A^3$$

where $\alpha > 0$ and F_{mn} is dependent on a projection of the nonlinear terms. The nonlinear Chuoke problem produced $F_{mn} > 0$, a destabilization, while the nonlinear Forscheimer's term is stabilizing when the flow is from the denser to the lighter fluid. The μ-terms in (18) do not produce $F_{mn} < 0$. To produce stability of the fingers new mechanisms must be studied [66,85].

Peters and Flock [76] have introduced a "wettability number" into the Chuoke analysis and have correlated experimental data with the number. Using a water-wet and an oil-wet number they obtained much information from dimensional analysis [98] and the linear stability analysis. Dimensional analysis often produces such illuminating results, and is used in many fields. In this case, the dimensional analysis produces the conclusion that the physics is not complete concerning the wettability phenomenon and more understanding and new equations are needed to explain these effects.

The mobility ratio m is an example of a dimensionless quantity which gives significant information about the flow properties in viscous fingering problems. Other dimensionless quantities which yield information in porous media flow include the Peclet number, the Reynolds number, the Froude number, the Bond number, the Mangioni number, and others.

The importance of dimensional analysis in the design of scale model experiments would be reason enough to consider this topic. In some cases, however, totally unexpected results can be gleaned from dimensional analysis. By introducing the concept of a wetting number, Peters and Flock [76] were able to extend the linear stability results of Chuoke et al. [7] to include the concept of wettability and correlate these ideas with an experiment. More analysis should give insight into the microscopic and non-linear effects included in this wetting number.

An important example of a closely related stability problem in a controlled setting, rather than the highly inhomogeneous porous media flow, is the work on the Hele-Shaw cell. Recently, McLean and Saffman [56] have calculated the finger formed in a Hele-Shaw cell with surface tension and found that the finger was unstable. Observations from experiments contradict this conclusion. Also, Vanden-Broeck [104] has shown that it is possible to have an infinite number of solutions to the Saffman-Taylor-McLean problem. Saffman [87] has added new insight into this problem. Park and Homsy [65] have developed a comprehensive theory of the Hele-Shaw cell which allows wetting of the walls, basically including local 3-D effects, and promises many new results. A better understanding of the Hele-Shaw problem will give insights which should aid in the solution of porous media problems.

Generalizing these concepts to flow in porous media, Jerauld et al. [43] use a traveling wave solution of permanent form [85] from which to perturb. In [43], equations for the stability parameters are calculated numerically, and in a "middle" parameter range, these parameters behave qualitatively like the Chuoke solution. A similar approach has been taken by Yortsos and Huang [113]. The obvious next step in this line of research is the incorporation of surface tension and the resulting finger growth. Perhaps asymptotic methods will illustrate how the various pieces fit together.

4. Global Model. When viscous fingering occurs, the finger wavelengths are generally much smaller than the typical computational cells which are used in a simulator, and therefore cannot be represented on a standard numerical grid. In order to correctly model the physics of the fingering phenomenon, one would have to add a prohibitive number of grid points around the moving front. If the fingering effect is not modeled, in general an overly optimistic value of fluid recovery is predicted. Realizing the inability to realistically model the physics of fingering with reasonable grids, a lumped-parameter approach, similar in principle to the averaging of the Navier-Stokes flow equations which was previously discussed, has been tried. The original attempt along these lines was due to Koval [44]. He utilized a global Buckley-Leverett approach. The Buckley-Leverett equations were derived to predict the immiscible one-dimensional effects of water and or gas displacement of oil in a reservoir. Solving the Buckley-Leverett equations does not predict anything about the fingering, but it does yield the displacing phase's saturation as a function of distance. Koval [44] in a landmark paper showed how to use the Buckley-Leverett equations to predict the movement of oil and water in miscible displacement processes. To review briefly the Buckley-Leverett equations, let s_o, s_w be the saturation of oil and water respectively, and A the cross-sectional area of the bed. We assume the bed is completely saturated,

(31)
$$s = s_w = 1 - s_o$$

and that the flow rate is a constant U,

(32)
$$u_o + u_w = U \ .$$

In addition to Darcy's Law, the conservation of mass of the two incompressible fluids, assuming constant density and porosity, yields

(33)
$$\phi \frac{\partial s_o}{\partial t} = -\frac{\partial u_o}{\partial x} \ , \ \phi \frac{\partial s_w}{\partial t} = -\frac{\partial u_w}{\partial x} \ .$$

By substituting Darcy's Law (1), (31), and (32) into (33) we obtain

(34)
$$\phi \frac{\partial s}{\partial t} + U \frac{\partial f}{\partial x} = \phi \frac{\partial s}{\partial t} + U \frac{\partial f}{\partial s}\frac{\partial s}{\partial x} = 0 \ ,$$

or

$$\frac{1}{V_{pi}} = \frac{\partial f}{\partial s} \ ,$$

where V_{pi} is the pore volumes injected and

(35)
$$f = \frac{1}{1 + \dfrac{k_o \mu_w}{k_w \mu_o}} = \frac{1}{1 + \left(\dfrac{1-s}{s}\right)\dfrac{1}{HE}} = \frac{ks}{1 + s(k-1)} \ .$$

By assuming no interaction between the oil and water, $k_o = k_1 s_o$, $k_w = k_1 s_w$, so $k_o/k_w = (1 - s)/s$. Also, $k = HE$ where H is a heterogeneity factor and E is the effective viscosity ratio. By (34),

(36)
$$\frac{1}{V_{pi}} = \frac{\partial f}{\partial s} = \frac{k}{[1 + (k - 1)s]^2}$$

and

(37)
$$s = \frac{f}{k - (k - 1)f}$$

which from (35) yields

(38)
$$V_{pi} = \frac{k}{[k - f(k - 1)]^2}$$

or

(39)
$$f = \frac{k - \left(\frac{k}{V_{pi}}\right)^{1/2}}{k - 1}$$

which is valid at or beyond breakthrough. The pore volumes at breakthrough occurs at $f = 0$ and is $V_{pi} = 1/k$, while when all of the oil is displaced, $f = 1$, and $V_{pi} = k$ in this case. The oil recovery in pore volumes is then

(40)
$$\text{Oil Recovered} = \frac{1}{k} + \int_{1/k}^{k} \frac{1 - (k - \left(\frac{k}{V_{pi}}\right)^{1/2})}{k - 1}\, dV_{pi} = \frac{2(kV_{pi})^{1/2} - 1 - V_{pi}}{k - 1} .$$

This formula correlates well with experimental data and indicates how to do this averaging for more complicated systems. Dougherty [16] has developed a more complete model utilizing the Koval procedure, but is a little more difficult to use.

Koval [44] attempted to account for the viscous fingering effect by adjusting the viscosity and/or density of the displacing phase when conditions were suitable for fingering to occur. Two variants of this approach are being used extensively in simulation: that of Koval [44] and that of Todd and Longstaff [101]. In Koval's approach, the effective viscosity for the injected fluid is calculated by

(41)
$$\mu_{eff} = \frac{\mu_o}{\left[0.78 + \left(\frac{\mu_o}{\mu_s}\right)^{1/4}\right]^4}$$

where μ_o and μ_s are the viscosities of the oil and the pure solvent, respectively. Todd and Longstaff modify both the viscosity and the density via a mixing parameter w [0,1] which is empirically matched to flooding data. Often, the results of simulation are quite sensitive to w which can be a function of grid size used.

Other attempts to incorporate the effects of fingering in a simulator without trying to model the physics on its smaller scale are discussed by Gardner and Ypma [26] and by Fishlock and Rodwell [24]. None of these attempts is completely satisfactory and the authors feel this is an important area for future research. Attempts to use statistics of the flow to empirically determine the longitudinal and transverse dispersion coefficients in the dispersion tensor (28) to incorporate fingering effects are in progress.

5. Conclusions. The concept of viscous fingering in porous media is a complex, and as of yet, not well understood phenomenon. Because of the direct impact of viscous fingering technology on improved and enhanced oil recovery, research on these problems is accelerating rapidly.

The physics and chemistry involved in the dynamic motion of the contact angle with both water-wet and oil-wet effects must be examined in more depth. These effects, coupled with nonlinear three-dimensional surface tension effects in a porous medium should be scrutinized for stabilization of the viscous fingers. Finger growth must be assessed by nonlinear stability techniques.

Finally, after the viscous fingering phenomenon is understood well enough to include finger growth, coalescence, etc., a statistical model using the ideas of Section 4 should be developed to improve the simulations of real oil reservoirs. The resulting predictive models should be amenable to modern control schemes which have the promise of significantly improving current production levels.

Acknowledgement. We would like to thank the participants in the 1983 Special Year in Energy-Related Mathematics at the University of Wyoming; in particular, we thank Aubrey Poore, Robert Heinemann, Don Peaceman, Hossein Kazemi, Larry Young, Joe Keller, John Buckmaster, and Jim Thomas.

REFERENCES

[1] J. BEAR, Dynamics of Fluid in Porous Media, American Elsevier, New York, 1972.

[2] J. BEAR and C. BRASESTER, On the flow of two immiscible fluids in fractured porous media, Proceedings First IAHR Symposium on Transport Phenomena in Porous Media, Haifa, 1969, pp. 177-207.

[3] A.L. BENHAM and R.W. OLSON, A model study of viscous fingering, Soc. Pet. Eng. J. (June 1963), pp. 138-144.

[4] B. BENJAMIN and F. URSEL, The stability of the plane free surface of a liquid in vertical periodic motion, Proc. Roy. Soc. A 255(1955), pp. 505-515.

[5] R.J. BLACKWELL, J.R. RAYNE, and W.M. TERRY, Factors influencing the efficiency of miscible displacements, Trans. AIME 216(1959), pp. 1-8.

[6] V. CASULLI and D. GREENSPAN, Numerical simulation of miscible and immiscible fluid in porous media, Soc. Pet. Eng. J. (October 1982), pp. 635-646.

[7] R.L. CHUOKE, P. VAN MEURS, and C. VAN DER POEL, The instabililty of slow, immiscible, viscous liquid-liquid displacements in permeable media, Trans. AIME 216(1959), pp. 188-194.

[8] E.L. CLARIDGE, A method for designing graded viscosity banks, Soc. Pet. Eng. J. (October 1978), pp. 315-324.

[9] P. CONCUS, Static menisci in a vertical right circular cylinder, J. Fluid Mech. 34(1968), pp. 481-95.

[10] G. DAGAN, Some aspects of heat and mass transfer in porous media, Proceedings First IAHR Symposium on Transport Phenomena in Porous Media, Haifa, 1969, pp. 55-64.

[11] S.H. DAVIS, Moving contact lines and rivulet instabilities - part 1 - the static rivulet, J. Fluid Mech. 98(1980), Part 2, pp. 225-242.

[12] S.H. DAVIS and G.M. HOMSEY, Energy stability theory for free surface problems: buoyance-thermocapillarity layers, J. Fluid Mech. 98(1980), Part 3, pp. 527-553.

[13] H. DEANS, Mathematical model for dispersion in the direction of flow in porous media, Soc. Pet. Eng. J. (March 1963), pp. 49-52.

[14] G. DE JOSSELIN DE JONG, The tensor character of the dispersion coefficient in anisotropic porous media, Proceedings First IAHR Symposium on Transport Phenomena in Porous Media, Haifa, 1969, pp. 259-267.

[15] G.P. DEMETRE, R.G. BENTSEN, and D.L. FLOCK, A multi-dimensional approach to scaled immiscible fluid displacement, J. Can. Pet. (July-August 1982), pp. 49-61.

[16] E.L. DOUGHERTY, Mathematical model of an unstable miscible displacement, Soc. Pet. Eng. J. (June 1963), pp. 155-163.

[17] F.A.L. DULLIEN, New network permeability model of porous media, AIChE J. 21(1975), pp. 299.

[18] J.M. DUMORE, Stability considerations in downward miscible displacements, Soc. Pet. Eng. J. (December 1964), pp. 356-362.

[19] E.B. DUSSAN V., On the spreading of liquids on solid surfaces: static and dynamic contact lines, Ann. Rev. Fluid Mech. 11(1979), pp. 371-400.

[20] E.B. DUSSAN V., The moving contact line: the slip boundary condition, J. Fluid Mech. 77(1976), pp. 665-84.

[21] E.B. DUSSAN V. and S.H. DAVIS, On the motion of fluid-fluid interface along a solid surface, J. Fluid Mech. 65(1974), pp. 71-95.

[22] D.C. DYSON, The energy principle in the stability of interfaces in Progress in Surface and Membrane Science, Academic Press, New York, 1978.

[23] R.E. EWING, Problems arising in the modeling of processes for hydrocarbon recovery, in The Mathematics of Reservoir Simulation, R.E. Ewing, ed., Frontiers in Applied Mathematics, Vol. 1, SIAM, Philadelphia, 1983, pp. 3-34.

[24] T.P. FISHLOCK and W.R. RODWELL, Improvements in the numerical simulation of carbon dioxide displacement, European Paris Conference, 1983.

[25] A.O. GARDNER, JR., D.W. PEACEMAN, and A.L. POZZI, JR., Numerical calculation of multidimensional miscible displacement by a method of characteristics, Soc. Pet. Eng. J. (March 1964), pp. 26-36.

[26] J.W. GARDNER and J.G.J. YPMA, Investigation of phase behavior-macroscopic bypassing interaction in CO_2 flooding, SPE 10686.

[27] R.M. GIORDANO and J.C. SLATTERY, Stability of static interfaces in a sinusoidal capillary, J. Colloid and Interface Sci. 92(1)(1983), pp. 13-24.

[28] J. GLIMM, D. MARCHESIN, and O. MCBRYAN, A numerical method for two-phase flow with an unstable interface, J. Comp. Phys. 39(1981), pp. 179-200.

[29] R.A. GREENKORN, J.E. MATAR, and R.C. SMITH, Two-phase flow in Hele-Shaw models, AIChE J. 13 (March 1967), pp. 273-279.

[30] S.P. GUPTA, J.E. VARNON, and R.A. GREENKORN, Viscous finger wavelength degeneration in Hele-Shaw models, Water Resources Research 9(1973), pp. 1039-1046.

[31] S.P. GUPTA and R.A. GREENKORN, An experimental study of immiscible displacement with an unfavorable mobility ratio in porous media, Water Resources Research 10(1974), pp. 371-74.

[32] B. HABERMAN, The efficiency of miscible displacement as a function of mobility ratio, Trans. AIME 219(1960), pp. 264.

[33] J.M. HAGOORT, Displacement stability of water drives in water-wet connate-water-bearing reservoirs, Soc. Pet. Eng. J. (February 1974), pp. 63-74.

[34] W.B. HAINES, Studies in the physical properties of soil, Part 5, J. Agricult. Sci. 20(1930), pp. 97-116.

[35] L.L. HANDY, An evaluation of diffusion effects in miscible displacement, Trans. AIME 216(1959), pp. 61.

[36] J. HELFFERICH, Theory of multicomponent, multiphase displacement in porous media, Soc. Pet. Eng. J. (February 1981), pp. 51-62.

[37] J.P. HELLER, The drying through the top surface of a vertical porous column, Soil Sci. Soc. Proc. 32(1968), pp. 778-86.

[38] J.P. HELLER, Onset of instability patterns between miscible fluids in porous media, J. Appl. Physics 37(1966), pp. 1566-1579.

[39] G. HETSRONI, Handbook of Multiphase Systems, McGraw-Hill, New York, 1982.

[40] C. HUH and L.E. SCRIVEN, Hydrodynamic model of steady movement of a solid/liquid/fluid contact line, J. Colloid Interface Sci. 35(1971), pp. 85-101.

[41] S. IRMAY, On the theoretical derivation of Darcy and Forscheimer formulas, Trans. Am. Geophysical Union 39(4)(1958), pp. 702-707.

[42] P. JACQUARD and P. SEGUIER, Mouvement de deux fluides en contact dans un milieu poreux, J. de Mecanique 1(1962), pp. 367-394.

[43] G.R. JERAULD, H.T. DAVIS, and L.E. SCRIVEN, Frontal structure and stability in immiscible displacement, SPE/DOE Fourth Symposium on Enhanced Oil Recovery 2(1984), pp. 135-144.

[44] E.J. KOVAL, A method for predicting the performance of unstable miscible displacement in heterogeneous media, Soc. Pet. Eng. J. (June 1963), pp. 145-154.

[45] C.R. KYLE and R.L. PERRINE, Experimental studies of miscible displacement instability, Soc. Pet. Eng. J. (September 1965), pp. 189-195.

[46] R.G. LARSON, L.E. SCRIVEN, and H.T. DAVIS, Percolation theory of two-phase flow in porous media, Chem. Eng. Sci. 36(1981), pp. 57-73.

[47] R.G. LARSON, H.T. DAVIS, and L.E. SCRIVEN, Displacement of residual nonwetting fluid from porous media, Chem. Eng. Sci. 36(1981), pp. 75-85.

[48] C.M. MARLE, Multiphase Flow in Porous Media, Institut Francais du Petrole Publications, Gulf Publishing Co., Houston, 1981.

[49] C.M. MARLE, On macroscopic equations governing multiphase flow with diffusion and chemical reactions in porous media in La Recuperation Du Petrole: Modeles Mathematiques Et Numerique, Vol.1, Racquencourt, 1982.

[50] MOREL-SEYTOUX, Introduction to flow of immiscible liquids in porous media, in Flow Through Porous Media, R. De Wiest, ed., Academic Press, 1969.

[51] J.C. MELROSE, Wettability as related to capillary action in porous media, Soc. Pet. Eng. J. 5(1965), pp. 259.

[52] J.C. MELROSE, Interfacial phenomena as related to oil recovery mechanisms, Can. J. Chem. Eng. 48(1970), pp. 638-644.

[53] D.H. MICHAEL, Meniscus stability, Ann. Rev. Fluid Mech. 13(1981), pp. 189-215.

[54] K. MOHANTY, H.T. DAVIS, and L.E. SCRIVEN, Physics of oil entrapment in water-wet rock, SPE 94406, Soc. Petrol. Engrg., 1980.

[55] N. MORROW, Interplay of capillary, viscous and buoyancy forces in the mobilization of residual oil, J. Can. Petrol. Technol. 18(3)(1979), pp. 35-46.

[56] J.W. MCLEAN, The Fingering Problem in Flow Through Porous Media, Ph.D. Dissertation, California Institute of Technology, 1980.

[57] J.W. MCLEAN and P.G. SAFFMAN, The effect of surface tension on the shape of fingers in a Hele-Shaw cell, J. Fluid Mech. 102(1981), pp. 455-469.

[58] A.H. NAYFEH, On the nonlinear Lamb-Taylor instability, J. Fluid Mech. 38(1969), Part 3, pp. 619-631.

[59] A.H. NAYFEH, Stability of liquid interfaces in porous media, Physics of Fluids 15(10)(1972), pp. 1751-1754.

[60] S.G. OH and J.C. SLATTERY, Interfacial tension required for significant displacement of residual oil, Soc. Pet. Eng. J. 19(1979), pp. 83.

[61] M.V. OSTROVSKY and R.M. OSTROVSKY, Dynamic interfacial tension in binary systems and spontaneous vs. pulsation of individual drops by their dissolution, J. Coll. and Interface Sci. 93(2)(1983), pp. 392-401.

[62] H.D. OUTMANS, Transient interfaces during immiscible liquid-liquid displacement in porous media, Soc. Pet. Eng. J. (June 1962), pp. 156-164.

[63] H.D. OUTMANS, Nonlinear theory for frontal stability and viscous fingering in porous media, Soc. Pet. Eng. J. (June 1962), pp. 165-176.

[64] H.D. OUTMANS, On unique solutions for steady-state fingering in a porous medium, J. Geophysical Research 68(1963), pp. 5735-5737.

[65] J.F. PADDAY, Theory of surface tension in Surface and Colloid Science, Vol. 1, E. Matijevic, ed., 1969, pp. 39-251.

[66] C.W. PARK and G.M. HOMSY, Two-phase displacement in Hele-Shaw cells: theory, J. Fluid Mech., to appear.

[67] A.C. PAYATAKES, C. TIEN, and R.M. TURIAN, A new model for granular porous media, AIChE J. 19(1973), pp. 67-76.

[68] D.W. PEACEMAN, Convection in fractured reservoirs -- the effect of matrix-fissure transfer on the instability of a density inversion in a vertical fissure, Soc. Pet. Eng. J. (October 1976), pp. 269-280.

[69] D.W. PEACEMAN and H.H. RACHFORD, JR., Numerical calculation of multidimensional miscible displacement, Soc. Pet. Eng. J. (December 1962), pp. 327-339.

[70] T.K. PERKINS and O.C. JOHNSTON, A review of diffusion and dispersion in porous media, Soc. Pet. Eng. J. (March 1963), pp. 70-84.

[71] T.K. PERKINS, O.C. JOHNSTON, and R.N. HOFFMAN, Mechanics of viscous fingering in miscible systems, Soc. Pet. Eng. J. (December 1965), pp. 301-317.

[72] T.K. PERKINS and O.C. JOHNSTON, A study of immiscible fingering in linear models, Soc. Pet. Eng. J. (March 1969), pp. 39-46.

[73] R.L. PERRINE and A.M. GAY, Unstable miscible flow in heterogeneous systems, Soc. Pet. Eng. J. (September 1966), pp. 228-238.

[74] R.L. PERRINE, Stability theory and its uses to optimize solvent recovery of oil, Soc. Pet. Eng. J. (March 1961), pp. 9-16.

[75] R.L. PERRINE, A unified theory for stable and unstable miscible displacement, Soc. Pet. Eng. J. (September 1963), pp. 205-213.

[76] E.J. PETERS and D.L. FLOCK, The onset of instability during two-phase immiscible displacement in porous media, Soc. Pet. Eng. J. (April 1981), pp. 249-258.

[77] E. PITTS, Penetration of fluid into a Hele-Shaw cell: the Saffman-Taylor experiment, J. Fluid Mech. 97(1980), pp. 53-64.

[78] H.S. PRICE, J.S. CAVENDISH, and R.S. VARGA, Numerical methods of higher-order accuracy for diffusion-convection equations, Soc. Pet. Eng. J. (September 1958), pp. 293-303.

[79] H.H. RACHFORD, JR., Numerical calculation of immiscible displacement by a moving reference point method, Soc. Pet. Eng. J. (June 1966), pp. 87-101.

[80] H.H. RACHFORD, JR., Instability in water flooding oil from water-wet porous media containing connate water, Soc. Pet. Eng. J. (June 1964), pp. 133-148.

[81] R. RAGHAVAN and S.S. MARSDEN, The stability of immiscible liquid layers in a porous medium, J. Fluid Mech. 48(1971), Part 1, pp. 143-159.

[82] R. RAGHAVAN and S.S. MARSDEN, A theoretical study of the instability in the parallel flow of immiscible liquids in a porous medium, Quart. Journ. Mech. and Applied Math. 26(1973), Part 2, pp. 205-216.

[83] J.G. RICHARDSON, Flow through porous media in Handbook of Dynamics, First Ed., Sect. 16, V. L. Streeter, ed., McGraw-Hill, New York, 1961, pp. 16.1-16.112.

[84] S. RICHARDSON, Some Hele-Shaw flows with time-dependent free boundaries, J. Fluid Mech. 1021(1981), pp. 263-278.

[85] E. RILLAERTS and P. JOOS, The dynamic contact angle, Chem. Eng. Sci. 35(1981), pp. 883-887.

[86] V.M. RIZHIK, I.A. CHARNI, and C. CHUNG-SIAN, Some exact solutions of the equations of unsteady two-phase fluid flow in porous media, Proc. Acad. Sci. USSR Div. Tech. Sci. Mech. and Mech. Eng. 1(1961), G.F. Teletzke, trans., pp. 1.

[87] P.G. SAFFMAN, Fingering in porous media, Lecture Notes in Physics 154, R. Burridge, ed., 1982, pp. 208-215.

[88] P.G. SAFFMAN and G. TAYLOR, The penetration of a fluid into a porous medium or Hele-Shaw cell containing a more viscous liquid, Proc. Roy. Soc. A 245(1958), pp. 312-329.

[89] P.G. SAFFMAN, Exact solutions for the growth of fingers from a flat interface between two fluids in a porous medium or Hele-Shaw cell, Quart. J. Mech. and Appl. Math. 12(1959), pp. 146-150.

[90] A.E. SCHEIDEGGER, General spectral theory for the onset instabilities in displacement processes in porous media, Geof. Pura Appl. 47(1960), pp. 41.

[91] A.E. SCHEIDEGGER, Growth of instabilities on displacement fronts in porous media, Physics of Fluids 3(1960), pp. 94.

[92] A.E. SCHEIDEGGER, On the stability of displacement fronts in porous media: a discussion of the Muskat-Aronofsky model, Can. J. Phys. 38(1960), pp. 153-162.

[93] A.E. SCHEIDEGGER, The Physics of Flow through Porous Media, University of Toronto Press, Toronto, 1960.

[94] A.E. SCHEIDEGGER, The statistical behavior of instabilities in displacement processes in porous media, Can. J. Phys. 39(1961), pp. 326-334.

[95] A.E. SCHEIDEGGER, General theory of dispersion in porous media, J. Geoph. Res. 66(1961), pp. 3273-3278.

[96] R.C. SHARMA and T.J.T. SPANOS, The instability of streaming fluids in a porous media, preprint.

[97] J.C. SLATTERY, Interfacial effects in the entrapment and displacement of residual oil, AIChE J. 20(1974), pp. 1145.

[98] J.C. SLATTERY, Momentum, Energy, and Mass Transfer in Continua, McGraw-Hill, New York, 1972; Second Ed., R.E. Krieger, Malabar, FL, 1981.

[99] J.C. SLATTERY, Interfacial transport phenomena, Chem. Eng. Commun. 4(1980), pp. 149-166.

[100] A.I. TAYLOR, The instability of liquid surfaces when accelerated in a direction perpendicular to their planes, Proc. Roy. Soc. London A 20(1950), pp. 192.

[101] M.R. TODD and W.J. LONGSTAFF, The development, testing and application of a numerical simulator for predicting miscible flood performance, Jour. Pet. Tech. 253(1972), pp. 874-882.

[102] P. VAN MEURS, Use of transparent three-dimensional models for studying the mechanism of flow processes in oil reservoirs, Trans. AIME 210(1957), pp. 295.

[103] P. VAN MEURS and C. VAN DER POEL, A theoretical description of water-drive processes involving viscous fingering, Trans. AIME 213(1958), pp. 103-112.

[104] J.M. VANDEN-BROECK, Fingers in a Hele-Shaw cell with surface tension, Physics of Fluids 26(8)(1983).

[105] J.E. VARNON and R.A. GREENKORN, Unstable two-fluid flow in a porous medium, Soc. Pet. Eng. J. (September 1969), pp. 293-300.

[106] N.C. WARDLAW, The effects of geometry wettability and interfacial tension on trapping in single pore-throat pairs, Journ. Can. Petrol. Tech. (May-June 1982), pp. 21-27.

[107] H.J. WELGE, A simplified method for computing oil recovery by gas or water drive, Pet. Abs. 195(1952), pp. 91-98.

[108] R.A. WOODING, Rayleigh instability of a thermal boundary layer flow through a porous medium, J. Fluid Mech. (1960), pp. 183-192.

[109] R.A. WOODING, The stability of an interface between miscible fluids in a porous medium, Zeit. Fur Ang. Math. Und Physik 13(1962), pp. 255-265.

[110] R.A. WOODING, Growth of fingers at an unstable diffusing interface in a porous medium or Hele-Shaw cell, J. Fluid Mech. 39(1969), pp. 477-495.

[111] R.A. WOODING, Instability of a viscous liquid variable density in a vertical Hele-Shaw cell, J. Fluid Mech. 7(1960), pp. 501-515.

[112] C.S. YIH, Dynamics of Nonhomogeneous Fluids, MacMillan Co., New York, 1965.

[113] Y.C. YORTSOS and A.B. HUANG, Linear stability of immiscible displacement including continuously changing mobility and capillary effects, SPE/DOE Fourth Symposium on Enhanced Oil Recovery 2(1984), pp. 145-162.

SIMULATION OF COMPOSITIONAL RECOVERY PROCESSES

ROBERT F. HEINEMANN*

Abstract. The relationships between the phase equilibria description in a compositional model and possible hydrocarbon recovery mechanisms are analyzed using the results from a one-dimensional model. Computations for both immiscible and miscible displacement problems are presented. The mass transfer mechanisms in these problems can be represented on a triangular phase diagram which we use to discuss restrictions placed on recovery by the reservoir composition and pressure and the injection gas composition. The miscible mechanisms are divided into vaporizing and condensing gas injection processes both of which have been postulated as the basis for carbon dioxide flooding.

1. Introduction. Reservoir simulators are the modern tools that engineers use to develop an optimal exploitation scheme for a given hydrocarbon reservoir. Black oil simulators have been the most widely used of these mathematical models and have studied conventional recovery techniques by assuming the hydrocarbon system consists of only two components, oil and gas. More recently, as the petroleum industry has proposed the use of tertiary processes and the development of deeper, more complex reservoirs, compositional simulators have been receiving considerable attention. These simulators treat the hydrocarbon as a multicomponent, multiphase mixture and use the principles of mass conservation and phase equilibria to track in time and space the pressure, saturations, and mole fractions of each component in each phase. The multicomponent nature of compositional models makes it ideal for accurately simulating a broad range of recovery processes such as: (a) miscible flooding by enriched hydrocarbon gas or CO_2, (b) cycling of gas-condensate reservoirs with dry gas, (c) injection of gas into volatile oil reservoirs, (d) natural depletion of volatile oil or gas condensate reservoirs.

Several modelling approaches to compositional problems appear in the literature. These computational efforts are categorized by their treatment of phase equilibria and by the degree of coupling between the thermodynamic constraints and fluid flow equations. The most popular algorithm has been an iterative scheme where the component material balances are used to form a pressure equation which is solved implicitly. The overall mole fractions and phase saturations are then obtained explicitly from their continuity equations. These compositions and pressure are input to a separate phase equilibria routine which yields the mole fractions and densities of each phase. The algorithm then returns to the pressure equation and the procedure is repeated. This

*Research and Development, Mobil Oil Corporation, DRD, P.O. Box 900, Dallas, Texas 75221

method is similar to IMPES approach in black oil simulators [1] and is first proposed in a compositional setting in the one-dimensional model of Roebuck et al. [2]. It has been applied to increasingly more complex problems in the works of MacDonald [3], Nolen [4], Van-Quy [5] and Kazemi et al. [6]. These papers are distinct from other IMPES formulations since they use K values and convergence pressures to perform the flash and phase equilibria calculations. Nghiem et al. [7] have extended this approach by employing a cubic equation of state in their phase equilibrium description. The Nghiem formulation also produces a pressure equation which can be linearized to give a symmetric, diagonally dominant Jacobian matrix. These desirable matrix features and the equation of state are the only features which differentiate this work from the earlier paper of Kazemi et al. [6].

Watts [8] has recently proposed a compositional model which uses a so-called sequential implicit algorithm that reportedly offers improved stability over the IMPES models. This formulation derives a set of pressure and saturation equations from the usual set of species continuity equations [9] and then solves this derived set using the well-known sequential algorithm from black oil simulators [10]. This model has an important advantage in that it naturally can solve both compositional and black oil problems with the same simulator.

The common algorithmic feature of the above formulations is that they are not based on Newton's method. They derive a pressure equation from physical principles and solve this equation implicitly by itself. This pressure is then used to solve for the remaining variables (saturations and mole fractions) in the formulation.

Coats [11] has developed a compositional model whose algorithm couples the phase equilibria and fluid flow equations in a fully implicit algorithm and solves these equations via Newton's method. Obviously with this scheme, the storage requirements and processing costs will be quite large and Coats' simulator is apparently not a field scale model like those of Kazemi et al. [6], Nghiem et al. [7] and Watts et al. [8].

A compositional model which may be considered a compromise of the above has been presented by Young and Stephenson [12] who clarified and extended an algorithm published by Fussell and Fussell [13] in 1979. Their formulation couples together all unknowns and equations in a Newton scheme with the important exception that the interblock transmissibilities are treated explicitly. While the explicit treatment of the transmissibilities may create a stability problem, this scheme will produce a Jacobian matrix which is much more sparse than the Jacobian of the Coats procedure. It has more coupling than the IMPES algorithm and uses this coupling to form an exact pressure equation through algebraic elimination in the Jacobian matrix. Young and Stephenson note that their linearization enables the Newton iterations to be considerably decreased from iterations counts reported by Nghiem et al. Furthermore, they indicate that by partial updating of the Jacobian in Newton's method their computational costs are competitive with those of the sequential algorithm. A more complete comparison of the work and storage requirements of each of these algorithms is the subject of another report [14] to which the reader is referred for more details. We have employed the Young and Stephenson scheme in this paper.

Compositional simulators may be used in a full field, engineering study of a given reservoir or in the analysis of a new recovery process applicable to a certain class of reservoirs. In the latter case, as exemplified in [15] and [16], one often attempts to determine the effect of the thermodynamic equilibrium on hydrocarbon recovery. To isolate the influence of equilibrium, the mass transfer occurring in the process is analyzed at a single set of conditions and totally divorced from the fluid flow in the reservoir. Hutchinson and Braun [17] and Stalkup [18] have presented excellent discussions on the relationships between phase equilibrium and recovery. These papers should be particularly useful to those familiar with the numerical analysis of fluid flow but inexperienced in the compositional analysis of mass transfer in reservoirs.

In this paper, we review the thermodynamic fundamentals in these two works and then present the mathematical description of both the phase equilibrium and fluid flow found in compositional models. We then present the recovery mechanisms described by Hutchinson and Braun and Stalkup and attempt to extend their analysis by showing how the thermodynamics and fluid flow interact to predict hydrocarbon movement. It should be noted that this paper is not written for the specialist in compositional simulation. Its purpose is to present the complexity of the physical processes analyzed by these simulators and to illustrate the utility of the modelling predictions with the hope of spurring interest in doing mathematical research on these problems.

2. Phase behavior fundamentals. The development and analysis of compositional recovery processes require a thorough understanding of the phase behavior of the hydrocarbon system. A triangular diagram is a convenient means of illustrating this behavior and is an accurate representation for three-component mixtures. Approximate diagrams for the methane-butane-decane system are shown in Figures 1 and 2. The phase behavior for multicomponent hydrocarbons is often also presented on a triangular diagram by combining the various components into three hydrocarbon mixtures. This diagram indicates the composition of each hydrocarbon phase at a specified pressure and temperature. Each corner of the equilateral triangle represents 100% of a given component. Inside the triangle, the perpendicular distances to the triangle's sides give the fractions of each component. The diagram also indicates the number of hydrocarbon phases present at the specified conditions. In Figure 1, a rounded phase boundary divides the diagram into two regions. Any mixture of the three components whose overall composition is inside the boundary will exhibit a distinct vapor phase that is in equilibrium with a distinct liquid phase. Tie lines connect the compositions of these phases such that a mixture with overall mole fractions, z_i , will have vapor and liquid compositions, x_i and y_i , given by the intersection of the tie line with the phase boundary. The three mole fractions are related by the following expression:

$$z_i = (1 - V)x_i + Vy_i \qquad i = 1,\ldots,N_c \tag{1}$$

where V, the fraction of hydrocarbon moles in the vapor phase, is measured as the distance between z_i and x_i divided by the distance between y_i and x_i .

The phase boundary in Figure 1 is also divided into two sections. The upper portion of this curve is termed the dewpoint curve and mixtures located on or above this curve exhibit a single gas phase with $V = 1$ and $z_i = y_i$. The lower section of the boundary is the bubble point curve and mixtures on or above the curve yield a single liquid phase with $V = 0$ and $z_i = x_i$. The dewpoint and bubble point curves are joined together at the critical point of the phase diagram. At this point, the liquid and gas become indistinguishable and any forces or tensions between the phases are removed. Hydrocarbon movement in the reservoir at this point is described by the flow of a single fluid phase described by linear relative permeability relationships and zero residual hydrocarbon saturations. Hence the goal of miscible flooding is the development of critical conditions in the reservoir and continued operation at or above this point.

Miscibility can be achieved in a reservoir generally in two ways. If the injection fluid is mixed with a reservoir fluid and the resulting mixture forms only a single phase, the fluids are said to be first contact miscible. From Figure 1, all mixtures along the methane-butane side and below the phase boundary are first contact miscible with any liquid phase mixture. When the injection and reservoir fluids form two phases upon mixing (for example, mixing pure methane with a fluid below the bubble point line), the fluids are immiscible. However, two fluids which are immiscible upon first contact may form a critical mixture in the reservoir due to mass transfer of individual components between phases. This process is called multiple contact miscibility and will be discussed further in a later section.

3. Mathematical model. The mathematical model describing the flow of several components through porous media consists of:

1. species continuity equations governing the conservation of mass of each component
2. Darcy's Law expressing the relationship between fluid velocities and pressure
3. thermodynamic equations describing phase equilibria
4. mole fraction and saturation constraints
5. an equation of state

The model is simplified by employing the following assumptions:

1. Capillary pressure and gravity are neglected.
2. Mass transfer between hydrocarbon phases is instantaneous compared with fluid flow phenomena.
3. No hydrocarbons are present in the aqueous phase.
4. The hydrocarbons comprise a single liquid and a single gas phase.
5. The porous medium is homogeneous and isotropic.

Under these assumptions, a species continuity equation for each of the N_c hydrocarbon components can be written as

$$(2)\quad \phi\,\frac{\partial(\rho_o x_i S_o + \rho_g y_i S_g)}{\partial t} = -\nabla\cdot(\rho_o x_i v_o + \rho_g y_i v_g) + Q_i\,, \qquad i = 1,\ldots,N_c\,.$$

The water equation is

$$(3)\quad \phi\,\frac{\partial \rho_w S_w}{\partial t} = -\nabla\cdot\rho_w v_w + Q_w\,.$$

The material balances indicate that the accumulation of component i at any point in the reservoir is equal to minus the divergence of the mass flux of the oil and gas phases multiplied by the mole fraction of each phase.

Darcy's Law expresses the velocity of the three phases in terms of the pressure gradient and phase mobilities (e.g., kk_{ro}/μ_o). They are considered equations of motion in this model and are written as [1]

$$(4)\quad v_o = -\frac{kk_{ro}}{\mu_o}\nabla P \qquad v_g = -\frac{kk_{rg}}{\mu_g}\nabla P \qquad v_w = -\frac{kk_{rw}}{\mu_w}\nabla P$$

Since it can be shown that $\rho_o x_i S_o + \rho_g y_i S_g = z_i(\rho_o S_o + \rho_g S_g)$ we can combine the continuity equation and Darcy's Law to give the following balances:

$$(5)\qquad \phi \frac{\partial F z_i}{\partial t} = \nabla \cdot \left[\frac{\rho_o k k_{ro}}{\mu_o} x_i + \frac{\rho_g k k_{rg}}{\mu_g} y_i \right] \nabla P + Q_i\,, \qquad i = 1,\ldots,N_c$$

where $F = \rho_o S_o + \rho_g S_g$. Likewise, the water balance becomes

$$(6)\qquad \phi \frac{\partial W}{\partial t} = \nabla \cdot \frac{\rho_w k k_{rw}}{\mu_w} \nabla P + Q_w$$

where $W = \rho_w S_w$.

The concept of fugacity is used to establish thermodynamic equilibrium constraints. The fugacity of a component in each phase is a measure of the escaping tendency or potential transfer of that component to another phase. If the fugacities are identical in two phases, there is no net interphase transfer so that the local equilibrium assumption can be expressed as

$$(7)\qquad f_{io} = f_{ig}\,, \qquad i = 1,\ldots,N_c\,.$$

Fugacity is a complex function of temperature, pressure and phase composition which is evaluated by substituting an equation of state such as the Peng-Robinson equation,

$$(8)\qquad z^3 - (1 - B)z^2 + (A - 3B^2 - 2B)z - (AB - B^2 - B^3) = 0$$

into the definition of the component fugacity

$$(9)\qquad RT\,\ln \frac{f_i}{P y_i} = -\int_{\infty}^{V} \left[\frac{\partial P}{\partial N_i} - \frac{RT}{V} \right] dV - RT\,\ln z\,.$$

Integration of Equation (8) gives for the vapor phase [7]

$$(10)\qquad \ln \frac{f_{ig}}{P y_i} = \frac{b_i}{b_g}(z_g - 1) - \ln(z_g - B_g) - A_g \frac{\Sigma\, y_i a_{ij}}{a_g} - \frac{b_i}{b_g} \ln\left[\frac{z_g + 2.414 B_g}{z_g - .414 B_g} \right].$$

An analogous expression for the liquid phase fugacity is written by substituting x_i for y_i and substituting the liquid phase subscript for the vapor phase subscript of each parameter. A detailed presentation of the z, b_i, B, w and A parameters are reported in [7].

The following molar and saturation constraints complete the mathematical description:

$$(11)\qquad \sum_{i=1}^{N_c} x_i = 1\,, \qquad \sum_{i=1}^{N_c} y_i = 1\,, \qquad \sum_{i=1}^{N_c} z_i = 1$$

and

$$(12)\qquad S_o + S_g + S_w = 1\,.$$

It is computationally convenient to rewrite the saturation constraint in terms of F and W:

(13) $$F\left[\frac{L}{\rho_o} + \frac{V}{\rho_g}\right] + \frac{W}{\rho_w} = 1$$

Also, the three mole fractions must be related by the overall balance written in Equation (1).

The following equation set then comprises our mathematical model:

	Number of Model Equations	Equation Number
Equilibria Constraints	N_c	7
Overall Molar Constraint	N_c	1
Species Balances	N_c	5
Water Balance	1	6
Molar Constraints	2	11
Saturation Constraints	1	12
	$3N_c + 4$	

The corresponding set of primary unknowns is:

	Number of Unknowns
x_i	N_c
y_i	N_c
z_i	N_c
P	1
F	1
W	1
V	1
	$3N_c + 4$

In the scheme of Young and Stephenson, Equations (1) and (11) are used to eliminate y_{N_c}, z_{N_c}, and all the x_i's from the list of unknowns. This algorithm then solves for $2N_c + 2$ unknowns at each grid block for every Newton iteration during the simulation. A thorough presentation of the Young and Stephenson procedure is given in their paper [12] and is not reviewed here.

4. Examples. We now present the results of solving the mathematical model to analyze two recovery processes in a one-dimensional reservoir. Gas was injected in both problems into a three-component oil with an initial composition of 44% methane, 11.5% butane, and 44.5% decane. The reservoir was initially at 1880 psi and 160° F with a porosity of 15% and a permeability of 20 md. The bubble point pressure of the in-place fluid was 1818 psi (V = 0) and the water was set to its irreducible value of 22%. The production at one end of the reservoir was fixed at 200 lb•moles/day, and in the first example, the injection at the other end was held to 154 lb•moles/day. Twenty grid blocks were used to discretize the reservoir.

Figure 3 presents the gas saturation as a function of normalized reservoir length at various times. This figure simply shows that gas appears by injecting a volatile component, methane, into the reservoir and illustrates the advance of the two-phase zone toward the production cell. Except for the appearance of a small amount of gas due to pressure decrease at the well, the reservoir produces only liquid until about 400 days.

After this breakthrough time, gas is produced at an increasing rate as shown by the relatively high gas saturations at 624 days.

Figure 4 illustrates the three methane mole fraction profiles in the reservoir. The methane profiles are quite similar to the gas saturation curves with the 184 day results indicating that the two-phase region has progressed to $x/L = 0.45$. Note that from $x/L = 0.45$ to 0.80 at this time, the overall and liquid phase mole fractions coincide since the reservoir is above the bubble point in this region. (The dashed vapor phase mole fractions drawn at these conditions represent the bubble point compositions which would be in equilibrium with the liquid phase at the bubble point pressure. Obviously the vapor phase is not present at the reservoir conditions.) We also point out that liquid and vapor profiles are nearly identical for the two sets of results shown and this indicates that the mass transfer between the phases is quite limited.

Figure 5 illustrates the variation of overall decane mole fraction with time. These profiles indicate that decane is simply being diluted from its initial value of 0.445 by the injection of methane. This injection tends to flatten the decane profiles particularly after the gas breakthrough since decane is the least volatile compound in the system and must dominate the liquid phase. The location of the two-phase zone is well defined by the composition plateaus in the 80, 184 and 400 day results. This occurrence has also been noted by Coats [2].

The limited mass transfer phenomena in this example can be explained by an analysis of the phase diagram [17,18]. An approximate phase diagram in Figure 6 contains an enlarged phase envelope for clarity of presentation. We also fix this boundary by assuming that the pressure (or at least the average pressure) in the reservoir is relatively constant. In this first example, we inject a gas G_1 (pure methane) into the reservoir liquid L_1. Our goal is to enrich this gas by forming mixtures which migrate to the critical point. When G_1 comes into contact with L_1, a mixture M_1 results and since this mixture is in the two-phase region, it separates into a gas G_2 and a liquid phase L_2. Continued injection of gas then pushes G_2 further into the reservoir where it contacts the original reservoir liquid L_1. Now a second mixing step can occur - G_2 contacts L_1 forming M_2 which splits into G_3 and L_3. G_3 is only slightly more enriched than G_2 in butane and decane compositions. When G_3 moves out and contacts L_1, a mixture M_3 is produced but now the resulting two phases have compositions which are nearly identical to those of G_3 and L_3. This indicates that the line mixing G_3 and L_1 together coincides with an equilibrium tie line so that no further enrichment can occur past the composition of G_3 as shown in the triangular diagram. The composition of the injection gas and the slope of the tie lines have prevented the enrichment of the lean gas in this example, and recovery here is attributed to the immiscible phenomena shown in Figures 3 through 5.

Figure 7 illustrates the changes in the example problem required to achieve miscible displacement by a multiple contacting process termed vaporizing gas drive [17,18]. We illustrate this process by first moving the reservoir liquid L_1 to the right thereby decreasing the decane mole fraction and increasing the butane mole fraction. Now consider the same mixing process as before - G_1 contacts L_1 forming M_1 which splits into G_2 and L_2. G_2 mixes with L_1 to give M_2 and G_3. G_3 now contacts L_1 to produce M_3, G_4 and L_4. We see now the enrichments of the gas phase (and liquid phase) are quite significant and move the system toward the critical point. Also note that the length of the "mixing" and tie lines decreases as the process occurs which indicates that the difference between the gas and liquid phases is becoming smaller. If we were to continue drawing the process on the phase diagram, these lines are eventually reduced to the single critical point, and miscibility is achieved in the reservoir.

Three distinct recovery processes can then be represented on this phase diagram by drawing a limiting tie line tangent to the critical point [17,18]. These are categorized by different injection gas and reservoir liquid compositions.

A. Gases with methane compositions between G_1 and G_{IN} along the 0% decane line can achieve miscibility with reservoir liquids to the right of tangent lines through multiple gas-liquid contacting. This is the vaporizing gas drive process explained above and is the basis for tertiary recovery in certain reservoirs (depending on their pressure and temperature) using CO_2 [15,16].

B. Gases with methane compositions between G_1 and G_{IN} can recover reservoir liquids to the left of the tangent line only by immiscible displacement as shown by our first computational example.

C. Gases to the right of G_{IN} can totally displace liquids to the right of the tangent since the two fluids are miscible without the multiple contact enrichment required in Case A. This outright miscibility process may require unfavorably large concentrations of the intermediate component in the injection gas, which would make this process uneconomic.

Another miscible flooding mechanism not discussed above is illustrated by the injection of a gas to the right of G_{IN} of Figure 7 into a reservoir liquid to the left of the tangent through the critical point. The process is called condensing gas drive and is applicable to reservoir liquids low in concentration of the intermediate components. The recovery mechanism is based on the condensation of intermediates out of the injection gas into the reservoir oil with multiple contacting eventually producing a critical fluid.

Figure 8 is used to illustrate this scheme [17,18]. The injection gas contacts the reservoir liquid and produces M_1, G_2 and L_2. As expected, L_2 is richer in the intermediate component than L_1 since butane is transferred from G_1. The gas G_2 has a lower viscosity and moves out into the reservoir leaving the liquid L_2 to be mixed with the injection gas G_1. The mixing of G_1 and L_2 produces M_2 which divides into G_3 and L_3. G_1 now contacts L_3 and this process continues until the equilibrium tie lines shrink into the critical point. If the phase diagram were drawn to proper scale, we would find the decane composition at the critical point is very low. Therefore, in a miscible flood, we expect to sweep out the decane, which is the desired product in this example.

Our second test problem exemplifies the condensing gas drive phenomena described above. This problem is identical with the first one except we now move the injection gas down the 0% decane line on the phase diagram by injecting 66 moles/day of butane with the 154 moles/day of methane.

Figure 9 presents the gas saturation profiles for this example. Recall at zero time no gas is present, but we see that as gas is injected, the gas saturations slowly increase and advance the interface of the two-phase zone into the reservoir. However, between days 186 and 290, a dramatic increase in S_g appears at the injection cell. This increase is caused by the transition from two-phase flow to single-phase flow as the reservoir produces critical conditions in the first block. We have arbitarily assigned the single phase to be the gas so that S_g at the critical point is 0.78. We were able to move the miscible conditions out into the reservoir without difficulty as evidenced by the results for 402, 506 and 620 days. The gas breaks through the production well between 402 and 506 days.

The methane composition profiles are shown in the next figure, and these profiles are obviously quite different from the profiles in the immiscible example. In this problem, the liquid and vapor mole fractions are quite different at the production end, but as one moves toward the injector, the mole fractions become increasingly closer together. At the critical point along these curves, the mole fractions become equal, $x_i = y_i = z_i$. The decreasing difference in phase composition is exactly consistent with this mechanism illustrated in Figure 8.

An interesting comparison can also be made between the decane profiles of our two example problems. Figure 11 presents the results for the condensing gas drive. For short times (< 85 days), the decane is again diluted in the first cells by the injection gas but later as miscible conditions develop, the decane profiles do not simply flatten out as in the earlier example. Now the decane is essentially swept out of the miscible cells, which is of course the goal of the condensing gas drive process.

Figure 12 presents the overall butane mole fractions in the reservoir which begin at 0.115 and change significantly as the 70% methane, 30% butane gas is injected. These profiles track the location of the miscible front; the overshoot which appears just downstream of the miscible front has been discovered by theoretical computations [2] and apparently confirmed by experimentation [5,6].

This paper has illustrated the utility of compositional simulators in analyzing hydrocarbon recovery processes. While the numerical methods employed in these simulators have not been emphasized here, improvement in the stability and accuracy of the current techniques is required. Well known problems in reservoir simulation such as grid orientation and numerical dispersion certainly arise in these models. Since the use of compositional models will surely increase, numerical research in this area seems certain to be an important topic for several years to come.

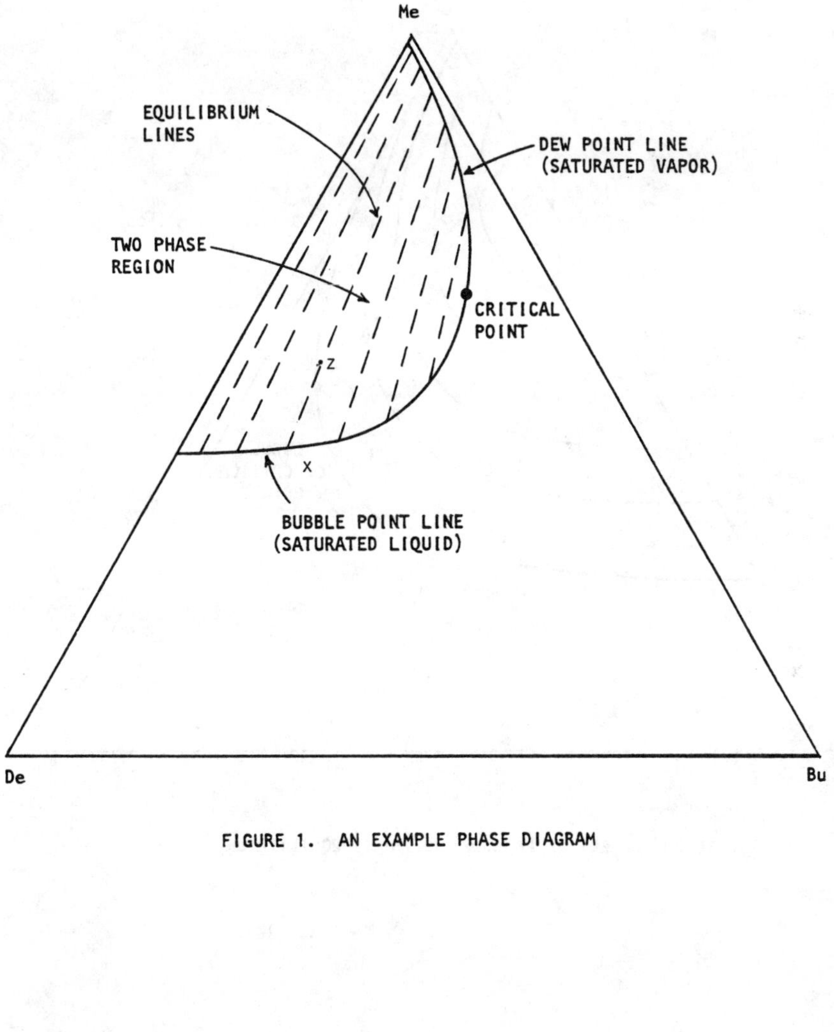

FIGURE 1. AN EXAMPLE PHASE DIAGRAM

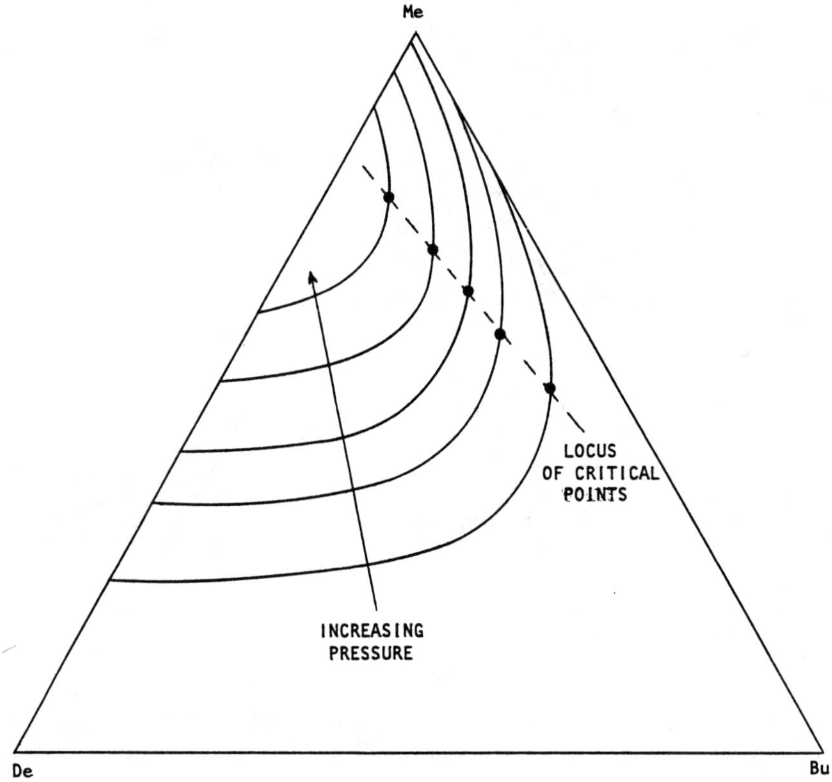

FIGURE 2. EFFECT OF PRESSURE ON PHASE EQUILIBRIUM

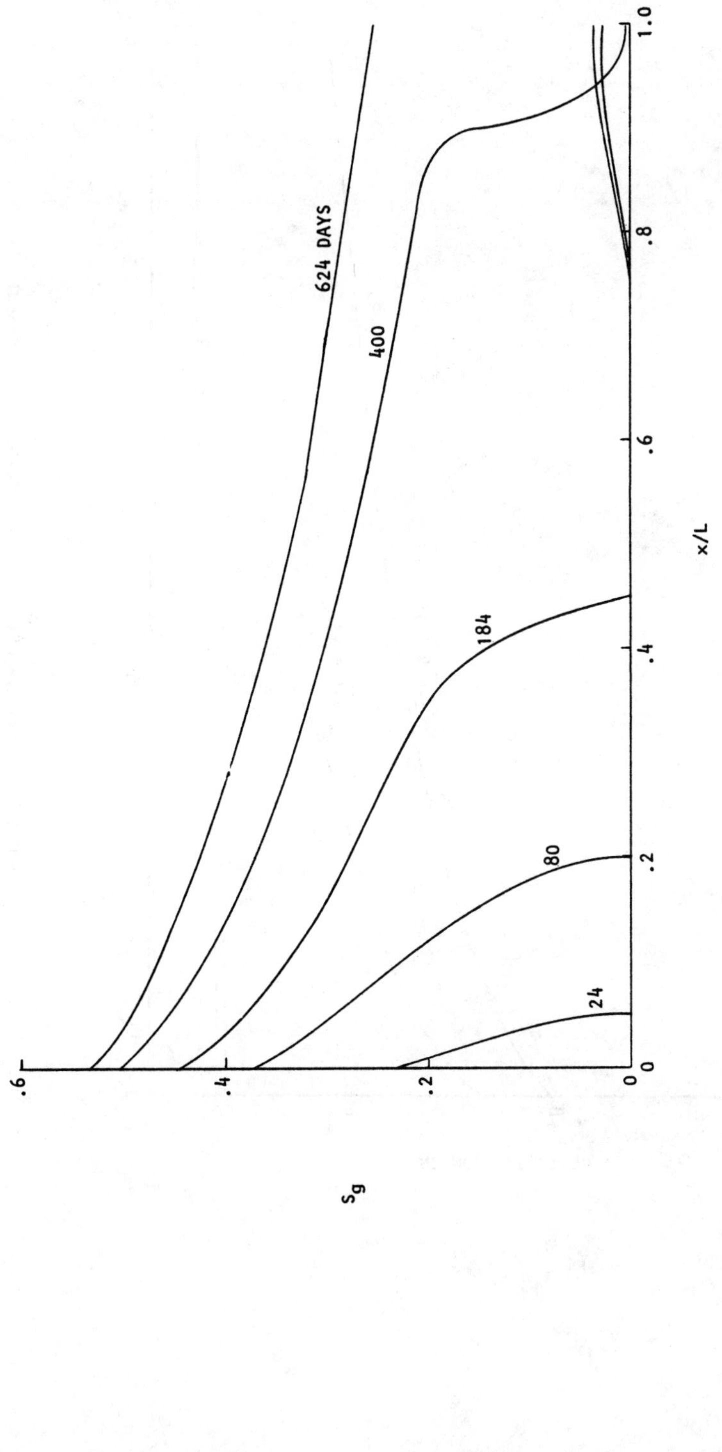

FIGURE 3. GAS SATURATION PROFILES FOR IMMISCIBLE DISPLACEMENT

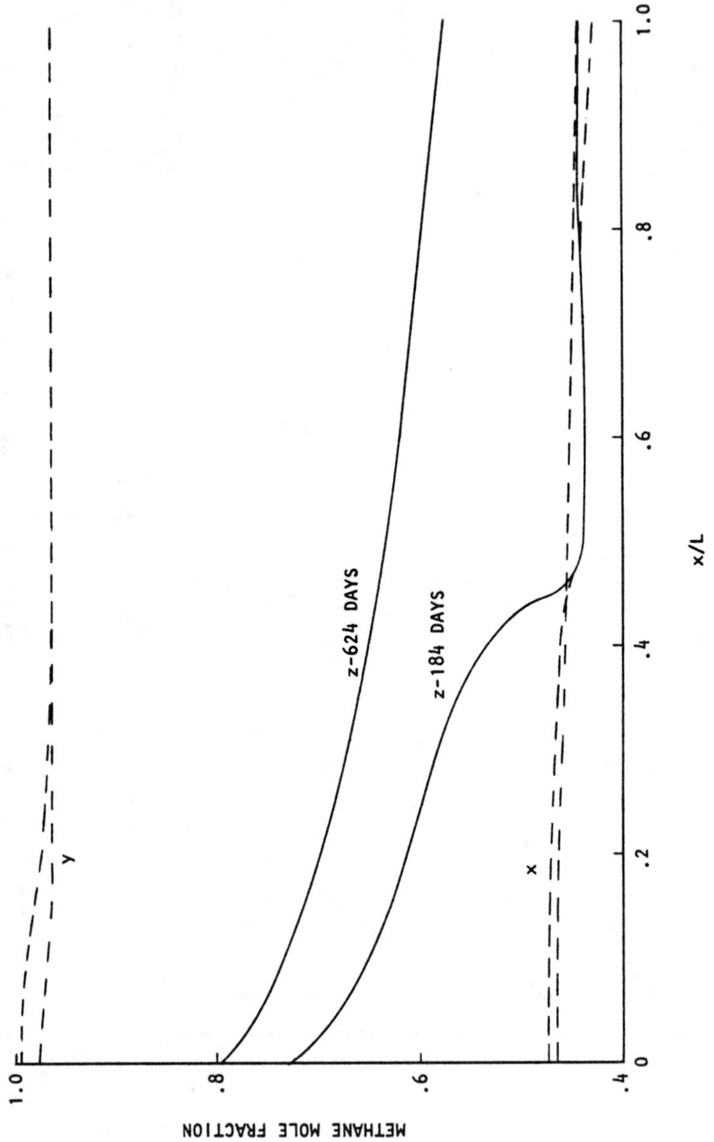

FIGURE 4. METHANE PROFILES FOR IMMISCIBLE DISPLACEMENT

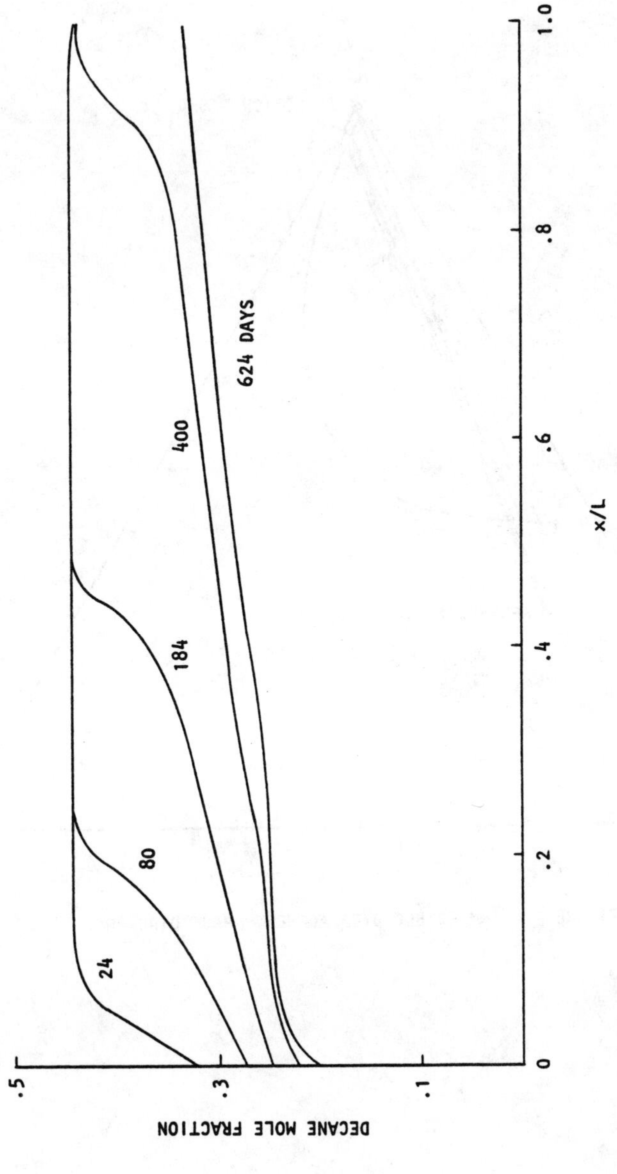

FIGURE 5. DECANE PROFILES FOR IMMISCIBLE DISPLACEMENT

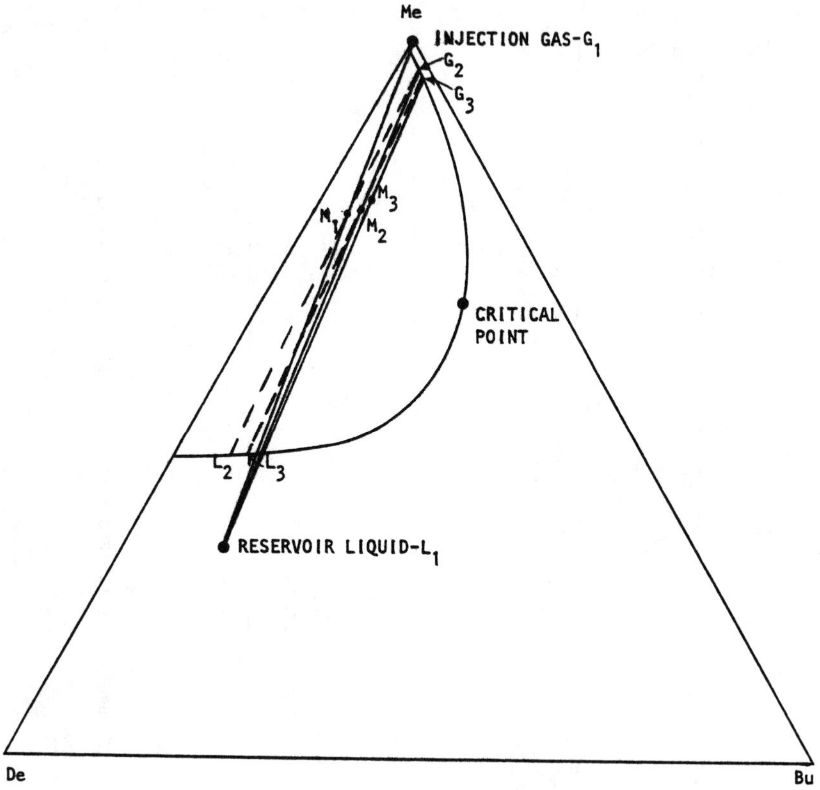

FIGURE 6. IMMISCIBLE DISPLACEMENT PHASE DIAGRAM

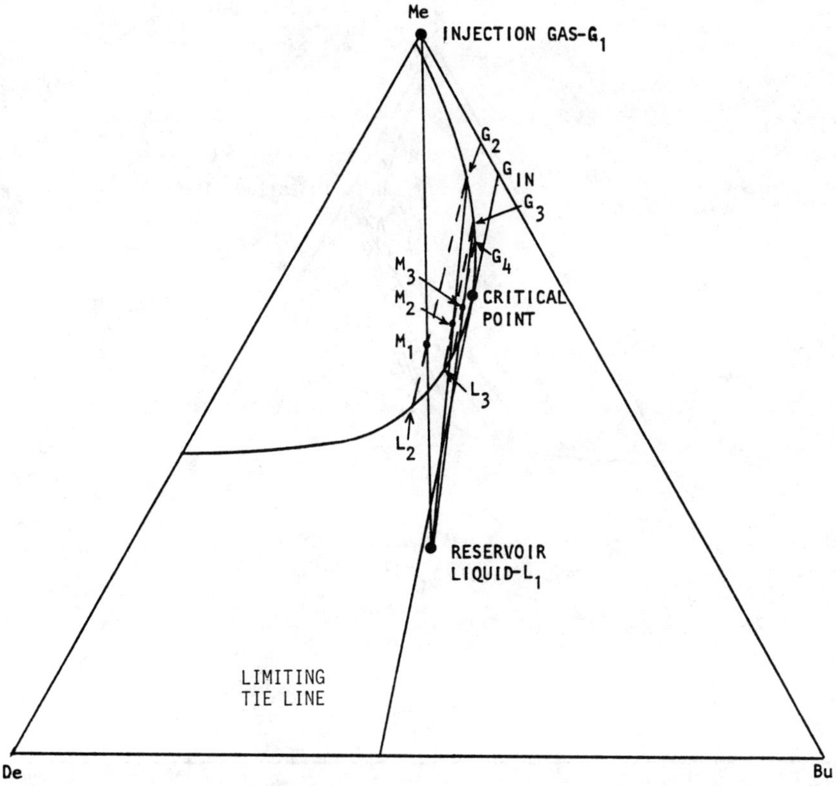

FIGURE 7. MISCIBLE DISPLACEMENT BY VAPORIZING GAS DRIVE

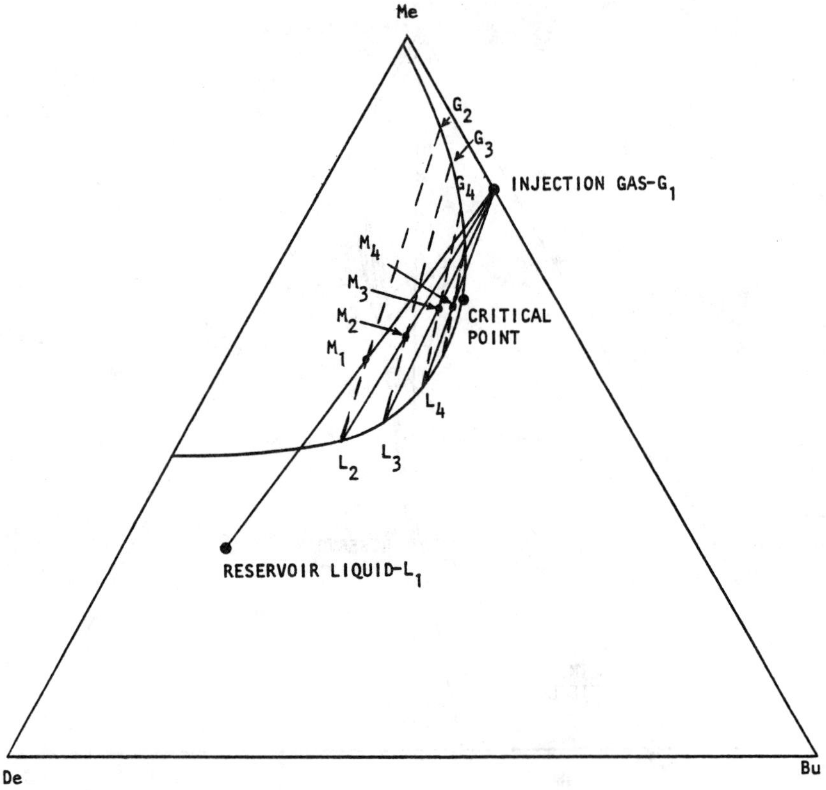

FIGURE 8. MISCIBLE DISPLACEMENT BY CONDENSING GAS DRIVE

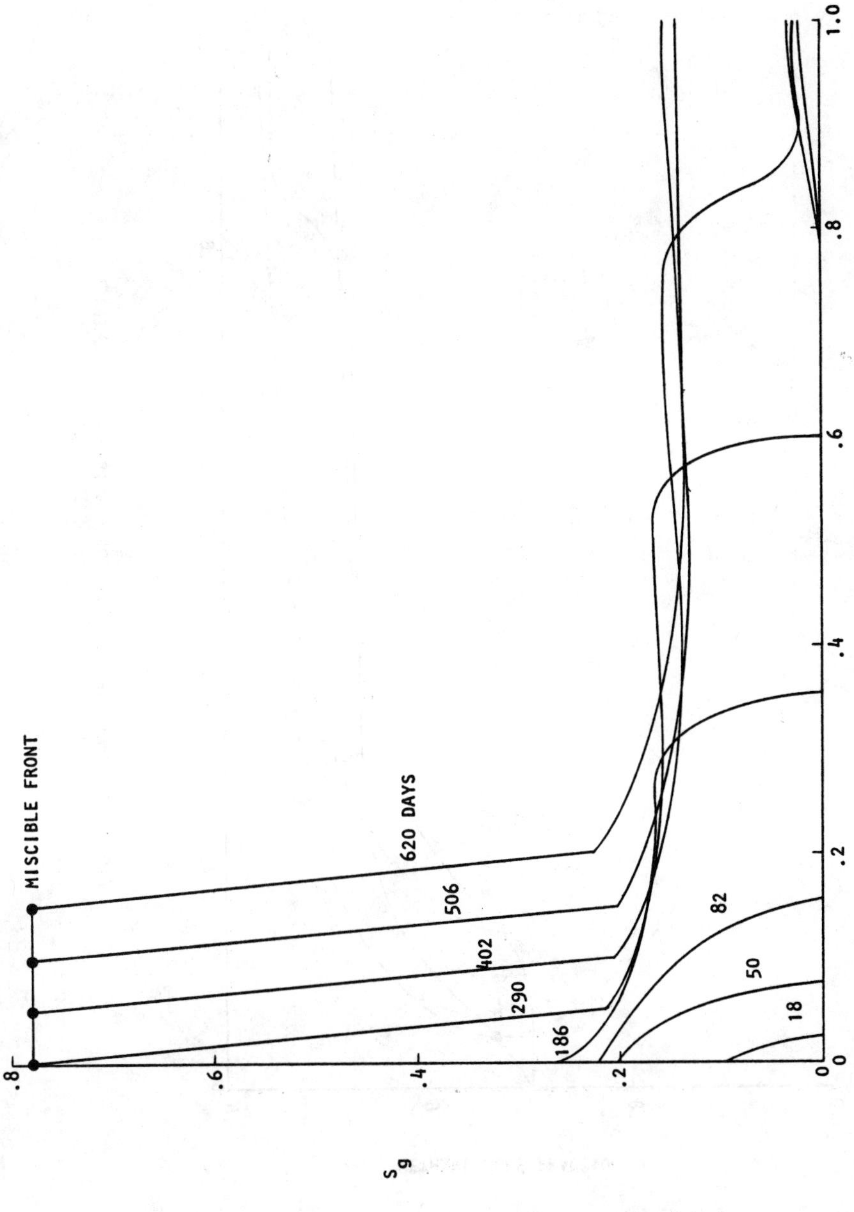

FIGURE 9. GAS SATURATION PROFILES FOR MISCIBLE DISPLACEMENT

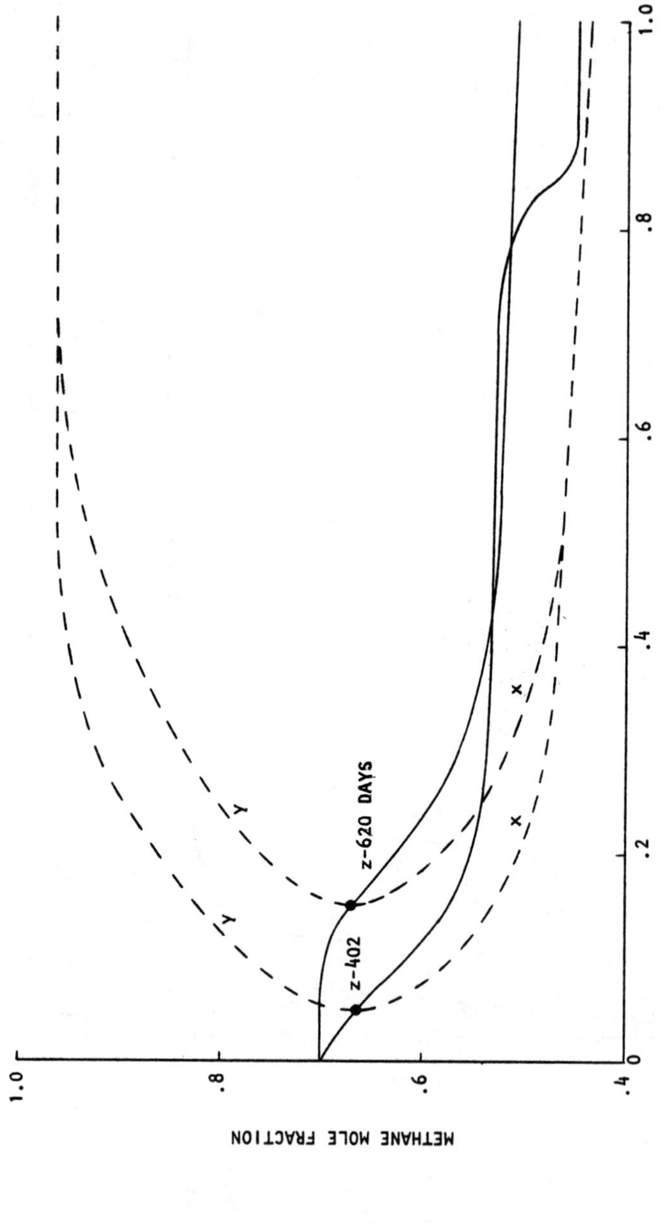

FIGURE 10. METHANE PROFILES FOR MISCIBLE DISPLACEMENT

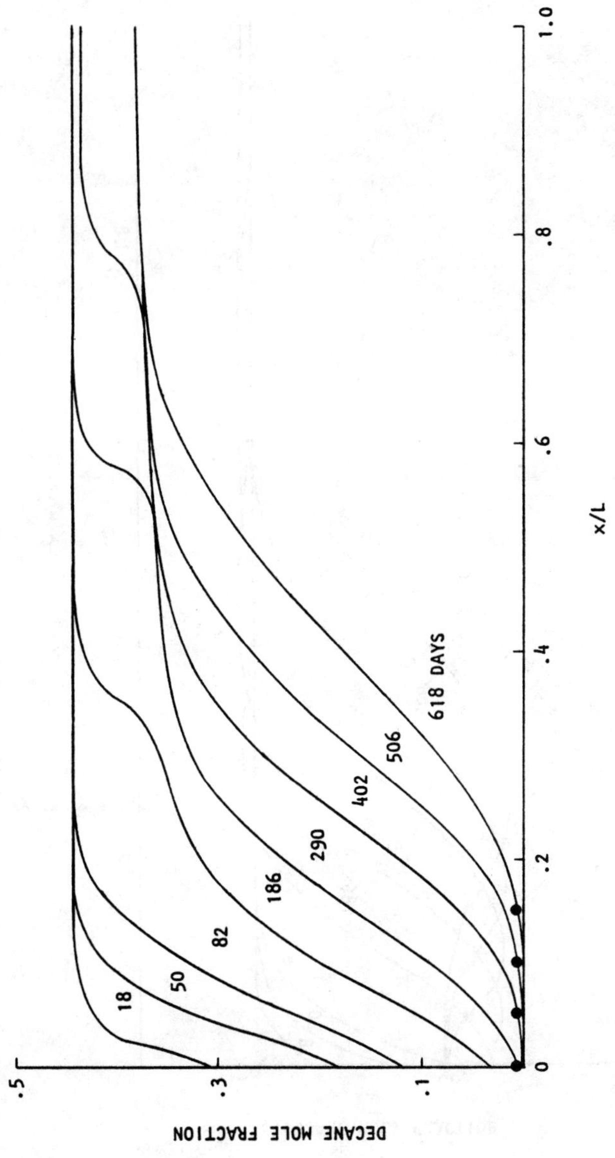

FIGURE 11. DECANE PROFILES FOR MISCIBLE DISPLACEMENT

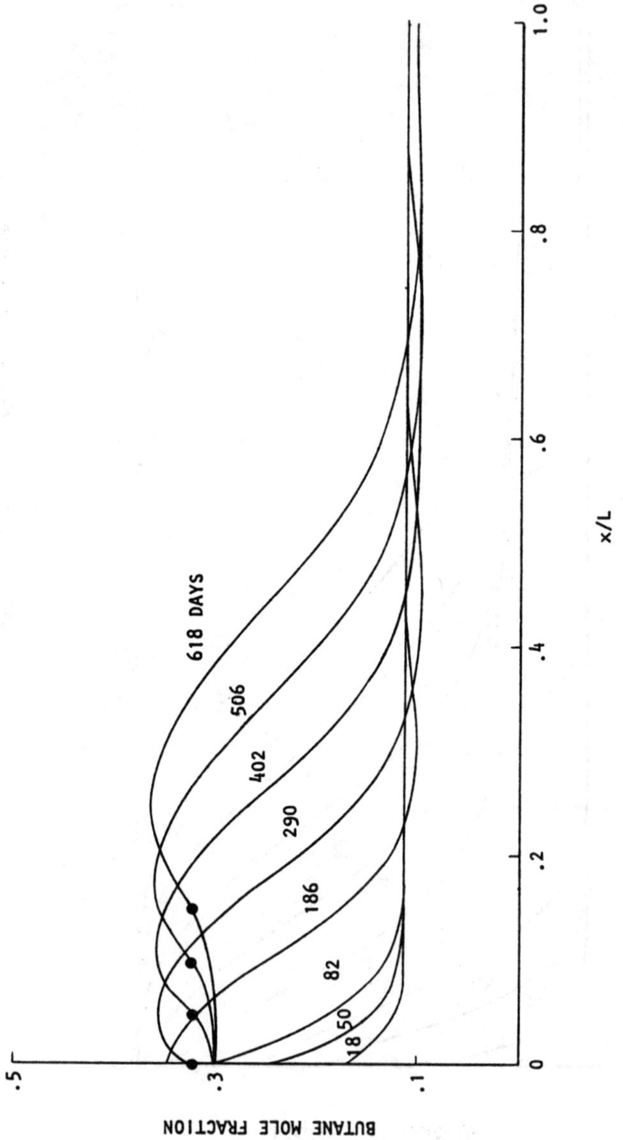

FIGURE 12. BUTANE PROFILES FOR MISCIBLE DISPLACEMENT

REFERENCES

[1] K. AZIZ and A. SETTARI, Petroleum Reservoir Simulation, Applied Science Publishers Ltd., London, 1979.

[2] I.F. ROEBUCK, G.E. HENDERSON, J. DOUGLAS, and W.T. FORD, The compositional reservoir simulator: case I - the linear model, Soc. Pet. Eng. J. (1969), p. 115.

[3] R.C. MACDONALD, Reservoir simulation with interphase mass transfer, Ph.D. Dissertation, University of Texas, Austin, 1971.

[4] J.S. NOLEN, Numerical simulation of compositional phenomena in petroleum reservoirs, SPE Paper 4274, Houston, 1973.

[5] N. VAN-QUY, P. SIMANDOUX, and J. CORTEVILLE, A numerical study of diphasic multicomponent flow, Soc. Pet. Eng. J. (1972), p. 171.

[6] H. KAZEMI, C.R. VESTAL, and G.D. SHANK, An efficient multicomponent numerical simulator, Soc. Pet. Eng. J. (1978), p. 355.

[7] L.X. NGHIEM, D.K. FONG, and K. AZIZ, Compositional modeling with an equation of state, Soc. Pet. Eng. J. (1981), p. 687.

[8] J.W. WATTS, A compositional formulation of the pressure and saturation equations, SPE Paper 12244, Seventh SPE Symposium on Reservoir Simulation, San Francisco, 1983.

[9] G. ACS, S. DOLESCHALL, and E. FARKAS, General purpose compositional model, SPE Paper 10515, Sixth SPE Symposium on Reservoir Simulation, New Orleans, 1982.

[10] A.G. SPILLETTE, J.G. HILLESTAD, and H.L. STONE, A high-stability sequential solution approach to reservoir simulation, SPE Paper 4542, 48th Annual Fall Meeting of the SPE, Las Vegas, 1973.

[11] K.H. COATS, An equation of state compositional model, Soc. Pet. Eng. J. (1980), p. 363.

[12] L.C. YOUNG, and R.E. STEPHENSON, A generalized compositional approach for reservoir simulation, SPE Paper 10516, Sixth SPE Symposium on Reservoir Simulation, New Orleans, 1982.

[13] L.T. FUSSELL and D.D. FUSSELL, An iterative technique for compositional reservoir model, Soc. Pet. Eng. J. (1979), p. 211.

[14] K.J. THELE, L.W. LAKE, and K. SEPEHRNOORI, A comparison of three equation-of-state compositional simulators, SPE Paper 12245, Seventh SPE Symposium on Reservoir Simulation, San Francisco, 1983.

[15] R.S. METCALFE and L. YARBOROUGH, The effect of phase equilibria on the CO_2 displacement mechanism, Soc. Pet. Eng. J. (1979), p. 242.

[16] M.P. LEACH and W.F. YELLIG, Compositional model studies - CO^2-oil displacement mechanisms, Soc. Pet. Eng. J. (1981), p. 89.

[17] C.A. HUTCHINSON, JR. and P.H. BRAUN, Phase relations of miscible displacement in oil recovery, AIChE J. (1961), pp. 7, 64.

[18] F.I. STALKUP, Miscible Displacement, SPE Monograph Volume 8, New York, 1983.

[1] K. AZIZ and A. SETTARI, Petroleum Reservoir Simulation, Applied Science Publishers Ltd., London, 1979.

[illegible]

SPECIAL YEAR IN ENERGY MATHEMATICS

PARTICIPANTS IN RESIDENCE
SPECIAL YEAR IN MATHEMATICS RELATED TO ENERGY

ALAIN BOURGEAT, Institut National des Sciences Appliquees de Lyon, 20 Av. A. Einstein, 69621 Villeurbanne, France

ROBERT BURRIDGE, Courant Institute of Mathematical Sciences, New York University, New York, New York 10012

JOHN F. EVERS, Department of Petroleum Engineering, University of Wyoming, Laramie, Wyoming 82071

RICHARD E. EWING, Mobil Field Research and Development Laboratories, Dallas, Texas 75221

Current Address: Departments of Mathematics and Petroleum Engineering, University of Wyoming, Laramie, Wyoming 82071

KEVIN P. FURLONG, Department of Geology and Geophysics, University of Wyoming, Laramie, Wyoming 82071

Current Address: Department of Geosciences, Pennsylvania State University-University Park, University Park, Pennsylvania 16802

JOHN H. GEORGE, Department of Mathematics, University of Wyoming, Laramie, Wyoming 82071

KENNETH I. GROSS, Department of Mathematics, University of Wyoming, Laramie, Wyoming 82071

HAL HUTCHINSON, Department of Petroleum Engineering, University of Wyoming, Laramie, Wyoming 82071

ELI L. ISAACSON, Department of Mathematics, University of Wyoming, Laramie, Wyoming 82071

JENG-ENG LIN, National Tsing Hua University, Applied Mathematics Institute, Hsinchu (300) Taiwan

Current Address: Department of Mathematics, George Mason University, Fairfax, Virginia 22030

AUBREY POORE, Mathematics Department, Colorado State University, Fort Collins, Colorado 80523

THOMAS RUSSELL, Denver Research Center, Marathon Oil Company, Littleton, Colorado 80160

SCOTT B. SMITHSON, Department of Geology and Geophysics, University of Wyoming, Laramie, Wyoming 82071

Professor Hutchinson died on August 13, 1983 in a tragic accident. The participants in the Special Energy Year acknowledge with great appreciation his enthusiastic contributions. The scientific community at the University of Wyoming keenly feels the loss of a strong colleague and friend.

SPECIAL LECTURE SERIES AND SUMMER SCHOOLS

ROCKY MOUNTAIN MATHEMATICS CONSORTIUM SUMMER SCHOOLS
University of Wyoming

July 1982
<u>Mathematical Seismology for Geophysical Prospecting</u>

Robert Burridge, Courant Institute of Mathematical Sciences, 251 Mercer Street, New York, New York 10012

Kenneth Larner, Vice President of Research and Development, Western Geophysical, P.O. Box 2469, Houston, Texas 77001

Robert Stolt, Director of Fundamental Research, Conoco Oil, Inc., P.O. Box 1267, Ponca City, Oklahoma 74601

June 1983
<u>The Mathematics of Large Scale Simulation</u>

Richard Ewing, Mobil Field Research Laboratories and University of Wyoming, Laramie, Wyoming 82071

Richard Kendall, J. S. Nolan and Associates, Inc., 11999 Katy Freeway, Suite 170, Houston, Texas 77079

Thomas Russell, Denver Research Center, Marathon Oil Company, Littleton, Colorado 80160

SPECIAL LECTURE SERIES ON ENERGY-RELATED MATHEMATICS

<u>Week of October 4, 1982</u>

DONALD PEACEMAN, Exxon Production Research Laboratory, P.O. Box 2189, Houston, Texas 77001

<u>Week of October 18, 1982</u>

JIM DOUGLAS, JR., Department of Mathematics, University of Chicago, Chicago, Illinois 60637

<u>Week of November 8, 1982</u>

G. D. SHANK, Denver Research Center, Marathon Oil Company, P.O. Box 269, Littleton, Colorado 80160

Week of December 6, 1982

NORMAN BLEISTEIN, Department of Mathematics and Computer Science, Denver University, Denver, Colorado 80210. Current Address: Department of Mathematics, Colorado School of Mines, Golden, Colorado 80401

ROBERT F. HEINEMANN, Research and Development, Mobil Oil Corporation, DRD, P.O. Box 900, Dallas, Texas 75221

Week of January 10, 1983

JOHN A. WHEELER, Project Leader, Geochemistry Section, Exxon Production Research Laboratory, Houston, Texas 77001

MARY WHEELER, Department of Mathematical Sciences, Rice University, Houston, Texas 77001

Week of January 17, 1983

HOSSEIN KAZEMI, Marathon Oil Company, P.O. Box 269, Littleton, Colorado 80160

OWE AXELSSON, Mathematisch Institute, Katholieke Universiteit, Toernooiveld, The Netherlands

Week of January 24, 1983

HAL HUTCHINSON, Department of Petroleum Engineering, University of Wyoming, Laramie, Wyoming 82071

Week of February 1, 1983

JOHN R. CANNON, Department of Mathematics, Washington State University, Pullman, Washington 99163

Week of February 7, 1983

JOSEPH KELLER, Department of Mathematics, Stanford University, Stanford, California 94305

Week of February 14, 1983

JOHN BUCKMASTER, Department of Mathematics and Applied Mechanics, University of Illinois, 311 Talbot Lab, Urbana, Illinois 61801

H. T. BANKS, Lefschetz Center for Dynamical Systems, Brown University, Providence, Rhode Island 02912

Week of February 21, 1983

EUSEBIUS DOEDEL, Department of Computer Science, Concordia University, 1455 De Maisonneuve Blvd. West, Montreal, Quebec H3G 1M8

Week of March 21, 1983

GUNTER H. MEYER, Department of Mathematics, Georgia Institute of Technology, Atlanta, Georgia 30332

Week of March 28, 1983

RICHARD KENDALL, J. S. Nolan and Associates, Inc., 11999 Katy Freeway, Suite 170, Houston, Texas 77079

Week of April 4, 1983

AZIZ ODEH, Mobil Research and Development Corporation, P.O. Box 900, Dallas, Texas 75221

Week of April 25, 1983

GEOFFREY S. S. LUDFORD, Mechanics Department, Cornell University, Ithaca, New York 14853

Week of May 11, 1983

L. C. YOUNG, Amoco Production Company, P.O. Box 591, Tulsa Oklahoma 74102